LC
40
.G34
2008

C

Homeschool
An American History

Milton Gaither

HOMESCHOOL
Copyright © Milton Gaither, 2008.

All rights reserved.

First published in 2008 by
PALGRAVE MACMILLAN™
175 Fifth Avenue, New York, N.Y. 10010 and
Houndmills, Basingstoke, Hampshire, England RG21 6XS
Companies and representatives throughout the world.

PALGRAVE MACMILLAN is the global academic imprint of the Palgrave Macmillan division of St. Martin's Press, LLC and of Palgrave Macmillan Ltd. Macmillan® is a registered trademark in the United States, United Kingdom and other countries. Palgrave is a registered trademark in the European Union and other countries.

ISBN-13: 978–0–230–60599–2 (hardback)
ISBN-10: 0–230–60599–0 (hardback)
ISBN-13: 978–0–230–60600–5 (paperback)
ISBN-10: 0–230–60600–8 (paperback)

Library of Congress Cataloging-in-Publication Data

Gaither, Milton.
   Homeschool : an American history / by Milton Gaither.
     p. cm.
     ISBN 0–230–60599–0 (alk. paper)—ISBN 0–230–60600–8 (alk. paper)
    1. Home schooling—United States—History. 2. Home schooling—Social aspects—United States. I. Title.

LC40.G34 2007
371.04'20973—dc22
                                            2007038055

A catalogue record for this book is available from the British Library.

Design by Newgen Imaging Systems (P) Ltd., Chennai, India.

First edition: July 2008

10 9 8 7 6 5 4 3 2 1

Transferred to Digital Printing 2012

For Rachel, Aidan, Susanna, and Macrina. Sequamini virtutes

Contents

Acknowledgments		ix
Introduction		1
One	The Family State, 1600–1776	7
Two	The Family Nation, 1776–1860	27
Three	The Eclipse of the Fireside, 1865–1930	51
Four	Why Homeschooling Happened, 1945–1990	83
Five	Three Homeschooling Pioneers	117
Six	The Changing of the Guard, 1983–1998	141
Seven	Making It Legal	175
Eight	Homeschooling and the Return of Domestic Education, 1998–2008	201
Notes		227
Index		265

Acknowledgments

This book would not have been possible without the willing assistance of many people. Several colleagues, including Sheila Moss, Jim Carper, Bob Hampel, Ed McClellan, Barbara Beatty, and John Fea, gave me valuable feedback on different parts of the manuscript. Thanks to the wonderful group of scholars in the History of Education Society whose work and comments at conferences inform even more of this book than the footnotes suggest. I benefited greatly from unpublished research and editorial assistance by Annie Blakeslee, Jesse McBride, Philip Martin, Autumn Carpenter, and Danielle DuBois. Messiah College's excellent library staff, especially Beth Mark and Dee Porterfield, located obscure references, answered questions, and filled countless interlibrary loan requests for me. Every homeschooling family I contacted was, without fail, generous with time and graciously answered my many questions. I cannot thank everyone with whom I spoke, but I am particularly grateful to Gregg Harris, Mark Hegener, Susan Richman, Linda Dobson, Cathy Duffy, John Holzmann, Jim Gustafson, and Deborah Stevenson. Special thanks to my wife and intellectual companion Elizabeth for enduring too many hours of table talk (and not a little pillow talk) about this book and giving me dozens of helpful leads and ideas over the past several years. Our children have grown up a lot as I've been writing it, and in its own way this book on home education has contributed its fair share to the learning going on all the time in our own kitchen and living room. I dedicate the book to them. Maybe they'll even read it!

Introduction

This book presents a history of education in the home in the United States. It is not just a history of the modern homeschool movement, though that movement plays a key role in the more recent history of education at home. It is a history of how the use of the home to educate children has changed and how it has remained the same from colonial times to the present. Taking such a long view allows us to correct some misconceptions about home education in the past and today. We need a book like this not least because advocates for and against homeschooling have often misrepresented the past in an effort to score political points. In the popular literature many historical misconceptions have circulated for so long that they have become commonplaces. Here are two of the most common.

On one hand there is a tendency among some to understand the modern homeschooling movement as a simple continuation of a process of education that has existed from time immemorial. The past is raided for examples of people, especially famous and influential people, who were taught at home. This virtuous and venerable tradition of home education is contrasted with the more recent and pernicious model of compulsory government schooling, and finally the modern homeschooling movement is introduced, described as a return to the original American (or in some cases Western) model of education. Here is an example of the sort of thing I'm talking about, coming from Theodore Forstmann, an advocate for parental choice in education:

> For the first 230 years of our history, parents, not government, were in charge.... Competence in reading, writing, and arithmetic was nearly universal at the time of the American Revolution. But by the mid-nineteenth century, a band of reformers led by Horace Mann of Massachusetts replaced our founding, free-market education system with a system of state-run education, with compulsory attendance and standardized curriculum.[1]

Now there is something salutary in this perspective. It reminds us that the historiography of education as it was written by public school leaders in the early twentieth century was simply wrong in its failure to attend to the family's role in education or to overstate the degree to which the American experiment nullified its reach. Though his own work later refuted this, a young Lawrence Cremin wrote in 1957 that "the European tradition of education centered in the family rather than in schools did not take root in the United States." Homeschoolers know better and have compiled an admirable list of historic individuals whose biographies collectively demonstrate the educative significance of the family. As the early chapters of this book will illustrate, the family was in fact the very center of education in early America.[2]

Yet this perspective, while accurately pointing out the anachronistic tendency of older histories of education to interpret the past as simply prolegomena to the public school, commits its own anachronism by reading current conflicts between family and state into the past. It doesn't often recognize that the modern homeschooling movement is in many ways fundamentally different from earlier efforts to educate children in the home. One of the central questions this book will be addressing is why and how education in the home shifted from being something that was done as a matter of course, actively encouraged by government, to being an act, even a movement, of self-conscious political protest against government.

A second tendency in some of the homeschooling historical writing that deserves scrutiny is the growing hagiography of the homeschooling movement's pioneers of the 1970s and 1980s. While different homeschoolers have different pantheons of saints (largely based on religious affiliation), all seem to share the assumption that homeschooling was brought back into prominence through the writings and works of great individuals. Frequently cited names include R. J. Rushdoony, John Holt, Raymond and Dorothy Moore, Gregg Harris, Michael Farris, and Mary Pride. No doubt these and many other leaders played crucial roles in making the modern homeschooling movement what it was and is. All will be covered in this book. But emphasis on the life and work of such notables tends to obscure the larger social forces at play during these decades that might go some distance in explaining not only why these thinkers and activists believed what they did but also why their views and initiatives met with such a strong grassroots response. It also misses the populist element, the reality that homeschooling's gains have come largely from the labors of a large group of ordinary Americans, almost all of them women.[3]

My task here then is twofold. First, I want to tell the history of education at home in the English colonies and the United States in a way that does not anachronistically valorize home education, pitting the home against the school or the family against the government. My tale must consider both the astonishing diversity of forms in which education in the home took place and the broader context of American social history that informed these practices. Secondly, I must explain the emergence since World War II of the self-conscious homeschooling movement proper, in a manner that goes beyond the "great man" history that has been typical of insider accounts, by attending to both the cultural climate in which the movement was incubated and the grassroots activism of thousands of Americans who participated.

Fortunately, historians have for several decades been producing excellent work in many fields that can help uncover some of these broader trends and themes. There are now well-developed literatures and entire subdisciplines of American history devoted to the history of childhood, the family, and education. Though much of these literatures are directly related to the topic of home education, until now no historian has explicitly made the connection. Professionals who spend their lives immersed in this historical world know that one of the most common complaints about the current state of affairs in American history is that there is so much good work being produced by so many scholars in such disparate fields that it has for some time been impossible to put it all together into one tidy synthesis. While there have certainly been intellectual currents of late that delight in the demise of older synthetic models of American history and look askance at any attempt to replace them, many professional historians do wish their work, often of the highest quality, could be made more accessible to the general reader. Good historical synthesis can play the role of bringing some of this high quality but often obscure historiography to more mainstream readers. That is one of the goals of this book.[4]

The book is arranged chronologically, with separate chapters covering the standard periods of American history from colonial times to the present. Chapter one describes home education in the British colonies. Chapter two does the same from the American Revolution to the Civil War. Chapter three covers the postwar period through the New Deal. Beginning with chapter four, the focus of the book narrows to the homeschooling movement. While other forms of education in the home certainly continued, the book does not dwell on such important topics as the use of tutors among the wealthy or for children with special needs. Chapter four lays out the broad cultural context within which the homeschooling movement was born. Chapter five looks in some detail at the work of three early leaders, each of whom made lasting contributions to the movement. Chapter six describes the development of the movement's infrastructure and organizations in all their complexity and controversy. Chapter seven comes to terms with the profound changes in homeschooling law the movement was able to secure. Chapter eight concludes with a look at recent developments in homeschooling, many of which are making homeschooling more and more like the domestic education of earlier centuries rather than a countercultural protest movement.

While sweeping generalizations are difficult to make for such a broad time span, I will hazard a few here. First, education in the home has indeed been a constant throughout the period, but its social meaning has changed dramatically. We will see, for example, a gradual shift from the colonial period when civil government aggressively *enforced* a certain sort of home education, to the slow and voluntary *eclipse* of home instruction by other institutions, then to the *antagonism* between home and school that has been a hallmark of the homeschooling movement, and finally to an increasing *hybridization* of home and school today. In my view, something truly revolutionary is happening with the homeschooling movement that can only be understood if we take the long view. Historian John Demos once wrote that the history of the family in America "has been a history of contraction and withdrawal; its central theme is the gradual surrender to other institutions of functions that once lay very much within the realm of family responsibility." But in our own time we are seeing a reversal of this longstanding pattern. Homeschooling is only the most obvious example here. Others include the rise in popularity of house churches among some conservative Protestants, the preference for live-in nannies and "au pairs" over day care among those who can afford it, the fashion of home-births assisted by midwives among many "crunchy" Americans, home-based hospice care, telecommuting, and the like. Some of the fuss over homeschooling may be due to the fact that it has been on the cutting edge of a larger renegotiation of the accepted boundaries between public and private, personal and institutional.[5]

Secondly, while concern for the moral and spiritual well-being of children has also been constant in our history, it has been addressed in many different ways. Here again, the modern homeschooling movement is perhaps only the most obvious example of a larger trend. Since the late nineteenth century, parents have looked up to experts and their bestselling child-care manuals (and institutions like schools) for help in raising their children. But historic deference to expertise has eroded dramatically in recent years, and a new spirit of self-reliance can be detected in such disparate phenomena as the rapid rise of do-it-yourself home improvement stores, self-diagnosis of medical conditions through Internet-based research, the valorization of independent film and music production, self-scanning checkout, online travel reservations, and so on. In a culture that mocks record company executives, second-guesses doctors, distrusts professional contractors, and delights in smart shopping, it is not surprising that many parents think of themselves as the most qualified arbiters of their children's moral and intellectual development.

Thirdly, different answers have been given throughout our history to the question of how to care for children without parents or whose parents are deemed unfit by others. Several strategies have been devised to accommodate orphans, abused children, and other "unfortunates," alternating between home-based remedies and institutionalization. Though it waxes and wanes, the home has regularly been called upon to educate other people's children in the past, and it may be called upon again in the future.

A final consistent theme throughout American history is the notion, held by almost everyone, that the fate of the nation rests on the strength of its families. But Americans have had quite different ideas about how to strengthen the family. It was concern for the family that inspired progressive reformers in the twentieth century to push for extended schooling, and the same commitment has inspired thousands more recently to reject that schooling. Homeschooling advocates believed that government regulation threatened the foundations of the family. Government representatives feared that unregulated families posed a threat to vulnerable children. All parties wanted stable families with happy, well-educated children; they just had different visions for how to get there.

For the professional historian, this book, especially the early chapters, may read too much like a summary of commonplaces readily available elsewhere. Specialists in the various fields in which I must dabble to tell this story will no doubt find some of what I say overly generalized and insufficiently complex. If there is one trend that can be said to characterize every field of American history, it is complexity. As historians probe more deeply into every period and region, using sources that include literary evidence, demographic and other quantitative data, and visual and material product, it has become perilous to hazard even the most tentative of generalizations about the past, for exceptions and counterexamples seem to sprout up everywhere. What I produce here, then, is my best effort to draw from this enormously complex "microhistorical" literature a coherent narrative of home education. I have no doubt that more could be said, and said better, by other historians, and I hope many of them will be inspired to improve on what I attempt here. I have done my best to produce a book that the professionals will find faithful and that everyone will find interesting. Whether I have succeeded in either task is for others to decide.[6]

Finally, a note on terminology. There has not yet emerged a standardized term to name the homeschooling movement. In this book I will render the words home and school as a compound only when describing the effort to educate children in the home as a deliberate rejection of and alternative to institutional schooling. Otherwise I will use word phrases like "domestic instruction," "home education," and so forth interchangeably. Given this, you might say that the central question this book addresses is how we got from "home school" to "homeschool" and back again. To find out we must begin with home school.

Chapter One

The Family State, 1600–1776

It wasn't that they didn't want to send their kids to school, but Amsterdam had a bit of a reputation even in 1608. In November 1607 the first flank of a group of Protestant separatists left their home in the village of Scrooby, England, for Amsterdam in hopes of finding enough religious toleration to allow them to follow God's revealed pattern of church government. They were called (rather derisively) "Brownists" after Robert Browne, who had achieved notoriety for his powerful tracts written in his youth against the established Church of England and in favor of the founding of separate congregations of the faithful without any episcopal oversight. Led by a few Cambridge intellectuals, they were mostly farmers ill suited to urban life in Holland. But Amsterdam was at the time one of the most tolerant places on earth, and there were already several hundred separatists there who had fled earlier persecutions, so it seemed a good idea at the time.[1]

But once the Scrooby separatists arrived, they were shocked to find not only their brand of separatism flourishing but also other sorts of religious sects: Calvinists of all varieties, Unitarians, Jews. Amsterdam was a thriving port city, and though religion saturated the place, the cosmopolitan air and Dutch culture were a bit of a shock to the Scrooby people, so much so that they feared for their children's futures. What was worse was that their own ranks were infected with innumerable theological quarrels, the most profound of which was that instigated by John Smyth, pastor of a group of English separatists who had relocated to Amsterdam some years earlier. Smyth's theology changed a lot over the course of his life, but when the Scrooby congregation arrived in Amsterdam, he had just hit on his most radical idea—that infant baptism was illegitimate and hence nobody was really a Christian, even members of separatist congregations. Smyth decided to baptize himself and then baptized several other adults in his congregation, but the theological innovation destroyed his church. Smyth himself later recanted from what he called this "damnable error," but the deed had been done and a new church, the Baptist church, was born. Anyway, this was the sort of thing going on in

Amsterdam when the farmers from the little village of Scrooby arrived. So they left.[2]

They resettled in the town of Leyden (now spelled Leiden), smaller and more rural than Amsterdam, which appealed to the faithful, and possessing a distinguished, though small, university, which appealed to their ministers. But here again theological controversy was raging, this time centered around a Leyden professor named Jacob Arminius, who rejected the Calvinist doctrine of predestination in favor of a view that God offered salvation to any who would choose to accept it. The leading Scrooby cleric, John Robinson, quickly got entangled in the dispute, becoming a vocal opponent of the Arminian position.

To make matters worse, not only was the theological air too full of opposing ideas to allow the Scrooby congregation to build the Kingdom of God in peace, there wasn't much of a living to be made in Leyden. With rare exceptions, few of these English emigrants knew any trade other than farming, so they were stuck with "hard and continual labor," as chronicler William Bradford put it. After twelve years of eking out a living in a foreign country doing menial tasks, these British farmers had had enough. Many of them were dead, many more had left the church, and they had made very few converts. Worse still, their own children had been forced into such heavy labors that, as Bradford wrote, "their bodies bowed under the weight...and became decrepit in their early youth." But most intolerable of all,

> many of their children, by these occasions and the great licentiousness of youth in that country, and the manifold temptations of the place, were drawn away by evil examples into extravagant and dangerous courses, getting the reins off their necks and departing from their parents. Some became soldiers, others took upon them far voyages by sea, and others some worse courses tending to dissoluteness and the danger of their souls, to the great grief of their parents and dishonour of God. So that they saw their posterity would be in danger to degenerate and be corrupted.[3]

Throughout their sojourn in Holland, the Scrooby families taught their children at home rather than send them to schools where they would learn Dutch grammar and manners. But even so, they were failing. So these Scrooby faithful decided at length to sail to the New World and to build a holy colony there. They became the Pilgrims. Generations of Americans have learned in elementary school of the Mayflower, Squanto, Thanksgiving, and the other tropes that make up the romance of Plymouth Colony, but it has not often been noted that one of the driving motivations behind the endeavor was the education of children.

For the first forty years of Plymouth Colony's existence there was no school at all. In the 1670s, a school was operated for a time, but by 1680 it had fizzled out. A similar situation existed in other colonies as well. Most learning occurred in the home, as mothers and fathers passed down values, manners, literacy, and vocational skills to their offspring. While this was done naturally, it was also a deliberate political philosophy. A Connecticut record from 1643 trenchantly summarizes the philosophy of education inspiring most early colonial settlements:

> The prosperity and well being of Comonweles doth much depend upon the well government and ordering of particular Families, which in an ordinary way cannot be expected when the rules of God are neglected in laying the foundations of a family state.

It is common knowledge that many British settlers moved to New England to build a holy commonwealth, a "city on a hill" that would shine its light on the darkness of British decadence. But what is less known is that the masonry being used to build this holy city was the family. The Pilgrims and many others came to the New World to build a family state.[4]

Inside the Family State

Of course the Pilgrims and other settlers were not the first North Americans to educate their children domestically. For millennia native tribes had been acculturating the next generation into their ancestral ways. For most tribes the family was subordinate to the community as a whole—the families were far more fluid in constitution than those of the Europeans, with frequent divorce, adoption of captured enemies into tribal families, and communal responsibility for discipline of children. One Naskapi native illustrated the general tenor by his derisive comment to a Jesuit missionary who was concerned that the easy social intercourse between Naskapi men and women might lead to illegitimacy, "Thou hast no sense" said the Naskapi man, "You French people love only your own children; but we love all the children of the tribe." Tribal child-rearing stressed discipline by public praise and shame rather than corporal punishment; the grooming of girls for sewing and farming, of boys for fishing and hunting. The frequent exogamy, or marrying outside the tribe, of native groups coupled with an absence of the institution of private property led to a general spirit of "informal sharing and reciprocity" that existed across tribes, making the preeminent social virtue generosity rather than acquisitiveness.[5]

But all of this changed with European contact. The informal social networks that had been built up for millennia and that relied on a delicate balance of natural resources and human cultural patterns were destroyed. By 1670 only about 10 percent of the original native population remained alive. Many were killed by European weaponry, but far more by European diseases. The natives who remained were alternately persecuted and driven from the land or adopted into white society. In 1619 Virginia passed a law requiring each town in the colony to take a certain number of Indian children into their homes so as to "advance their civilization." This same approach was taken for centuries as Protestant and Catholic missionaries together with government sought some way of breaking natives from their tribal ways and integrating them into Anglo-American life. The family and its domestic education were the front line of this acculturation project.[6]

But what sort of families did colonists actually have? A lot of ink has been spilled on this question. It used to be thought that colonists first reproduced the large extended families (several generations living under one roof) they had left behind in Europe, but that the pressures of industrialization and urbanization gradually whittled the American family down to its nuclear form—a father, mother, and their biological children. As historians found more and more evidence of the nuclear pattern existing earlier and earlier in American history, it became fashionable to argue that the large extended families of Europe failed in the New World due to the harsh wilderness environment. But historians studying Europe found that the nuclear pattern existed there as well, so it is now generally agreed that the dominant family pattern in North America (and early modern Europe for that matter) has always been nuclear and that colonial families were often *more* stable than those left behind in Europe.[7]

Several misconceptions about colonial families persist in the public imagination. It might surprise many to learn that men, on average, married around age twenty-five, and that most women they married were at least twenty. Though the number varied greatly by region and generation, families were not as large as one might expect. Women typically had between four and seven children in the seventeenth century, with that number dropping somewhat by the eighteenth century.

Though death during childbirth was a serious possibility, historians estimate that only 1 in 30 births proved fatal to the mother. Again, though child mortality was a real concern, three out of four children lived to adulthood. Children were usually spaced at least two years apart, perhaps in part due to the contraceptive effect of extended nursing. Though remarriage after the death of a spouse was standard practice, most colonists married only once and very few had a third marriage. Now, all of these claims are the most tentative of generalizations based on best guesses from spotty baptismal records, wills, deeds, tombstone evidence, and various literary remains, and there are of course many exceptions. Samuel Fuller, for example, died leaving nine people in his household: his wife, son, nephew, two servants, a ward, and two children who had been sent to him for education. Benjamin Franklin, though his own writings on the topic of the declining birth rate in the colonies is often cited as evidence for some of the claims made above, himself was the tenth son in a family of seventeen children born to one man by two wives. And many of the wealthier colonial families did in fact seek to replicate the more extended household of the British aristocracy, especially on Southern plantations. Virginia planter Captain Samuel Mathews' household, for example, included among his host of domiciled servants and slaves, eight shoemakers and their families.[8]

But most colonists were not wealthy. Most, in fact, lived in very cramped quarters, often only a single room with a loft perhaps. Everyone slept in one or two bedchambers, often three or four to a bed—adults and children, servants and guests. Since almost all colonists, even shopkeepers, were farmers, daily life in and around this small home was largely occupied by agriculture, horticulture, and food preparation and storage, thus making daily life "a continuous general apprenticeship in the diverse arts of living" for colonial children. Yet despite the difficult living conditions, which must have become even more suffocating during long New England winters, colonial records show remarkably little evidence of familial strife. If court records are an accurate guide here, conflict occurred much more frequently between neighbors (especially over property boundaries) than between spouses or children.[9]

Perhaps this was due in part to the absolute practical necessity of harmony for survival, especially in the early years of settlement. Early colonists faced the threat of the untamed wilderness without the institutional aides they left behind in Europe. If the family had traditionally played the lead role in the religious and educational life of children in Europe, in the New World the family often played the only role.

Thus for many colonists and their governments, the success of the colony depended primarily on the success of its families. Successive colonies passed law after law requiring parents to educate their offspring. In 1642 Massachusetts mandated that local officials called selectmen

> take account from time to time of all parents and masters, and of their children, especially of their ability to read and understand the principles of religion and the capital laws of this country.

A similar law was passed in Connecticut in 1650, in New Haven in 1655, in New York in 1665, and in Plymouth in 1671. In 1683 Pennsylvania issued an ordinance declaring that all parents and guardians

> Shall cause such to be instructed in reading and writing, so that they may be able to read the Scriptures and to write by the time they attain to twelve years of age; and that then they be taught some useful trade or skill.[10]

Most of these laws imposed fines or worse if parents failed in their duties. In Virginia, a fine of five hundred pounds of tobacco was to be levied when clergy discovered parents who were "delinquents in the catechizing the youth," though there is no evidence that anyone was ever actually fined. Perhaps no fines were levied because families were largely doing their appointed job. Failure to do so would have been an economic disaster, for the contribution of children to the domestic economy was crucial for the survival of all. It was the household that provided both the substance for a living and the education for making that living possible. While the roles of male and female were more fluid than they would later become, in general girls were taught food processing, manufacture of clothing and bedding, candle-making, brewing, and other domestic tasks from their mothers or older siblings in the household while boys picked up farming skills from their fathers and brothers.[11]

The family was the crucial institution for nearly all social services in the colonies. Each household was more or less a self-sufficient economic unit, a school, a vocational institute, a church and Sabbath school, a house of correction, and a welfare institution. Families were paid by town government to take in cripples and others who could not support themselves. Unmarried adults were in most places required by law to live in a household headed by a married couple. A New

Hampshire court in 1672, for example, ruled that a man who "lay in a house by himself contrary to the law of the country" must "settle himself in some orderly family in the town." As such examples show, it was not as if the family was an independent, autonomous agent. Colonial families were very tightly regulated by colonial government. In 1636, for example, Plymouth Colony decreed that only men approved by proper authority would be allowed "to be housekeepers or build any cottages." Connecticut forbade gambling in households in 1657. Virginian household heads were required by law to lead their families in daily prayer, to plant certain crops, to report annually the number of taxable persons in their households, and even to bring guns to church.[12]

Responsibility for executing this governance in many colonies fell on a vast array of petty officers whose chief task was to monitor families. In 1675 the Massachusetts General Court established "tithingmen" to monitor parental instruction and bolster good household government by reporting unruly children and adults. Each tithingman was appointed by the selectmen to "diligently inspect" ten or more families in a given neighborhood and report any unacceptable behavior. If fathers failed to maintain "well-ordered" families, the community asserted its responsibility for enforcing morality by some sort of intervention, even to the point of forcibly removing what one early law called "rude, stubborn, and unruly" children from their parents. And what is most remarkable, given the subsequent American valuation of privacy, freedom, and autonomy, most colonists seemed to accept this tight oversight, at least formally. Here, for example, is a selection from John Cotton's popular Catechism, memorized in many colonial households (1646):

> *Question*: What is the Fifth Commandment?
> *Answer*: Honor thy father and thy mother, that the days may be long in the land which the Lord thy God has given thee
> *Question*: Who are here meant by father and mother?
> *Answer*: All our superiors, whether in family, school, Church, and commonwealth[13]

From modern perspectives, this symbiosis between family and state seems not only invasive on civil liberties grounds, but all but incomprehensible. But if one understands the political orientation of most colonists the family state makes perfect sense. Historian Mary Beth Norton has marvelously explained for us how most colonists possessed what she calls a "Filmerian outlook" (named for English theorist Sir Robert Filmer) on family and government. Filmer simply codified what was the gut-level instinct of most premodern Europeans. For him, and for most colonists, the family and the state were analogous institutions, both created by God and grounded in a natural law of patriarchal authority. As the King governed by Divine right, so did the father. And both family and state were created by God to serve the same purpose—the peaceable government of society according to Divine law. But over time this unified understanding of both family and state broke down. The breakdown of the state is writ large in the revolutions in English political life in the seventeenth and eighteenth centuries. But in the family a similar revolution gradually took place. As the government notions of divine sanction for an eternal order gave way to contractarian explanations of the origin of political power, so in the family a more provisional, social contract view of family relations gradually took hold, symbolized for professor Norton in the rising popularity of John Locke's treatises on the family and its education.[14]

We will return to the Lockean orientation shortly, but for now the main point to note is that under this Filmerian unification of home and state, what was happening in colonial North America was the aggressive *enforcement* of home instruction by the government. A Connecticut court record from 1665 captured it well:

> Whereas reading the Scripture, catechizing of children and daily prayer with giving of thanks is part of God's worship and the homage due to him, to be attended conscientiously by every Christian family to distinguish them from the heathen who call not upon God, and the neglect of it a great sin... this court do solemnly recommend it to the ministry in all places, to look into the state of such families, convince them of and instruct them in their duty... but if any heads or governors of such families shall be obstinate and refractory and will not be reformed, that the grand jury present such person to the county court to be fined or punished or bound to good behavior, according to the demerits of the case.

And it is clear that the brunt of the responsibility for this godly upbringing was placed upon the father.[15]

The Last Days of the Patriarch

One of the most fascinating themes in American family history is the gradual shift in responsibility from father to mother for the religious and moral training of children. In early colonial society it was all dad. Laws were written for fathers and held fathers responsible for breaches in family order and education. Until the middle of the eighteenth century child-rearing manuals were addressed not to mothers but to fathers. In Puritan sermons, notes historian N. Ray Hiner, "hardly any attention was devoted to mothers in seventeenth-century recitations of parental duties...If the Puritan father had a relatively equal teaching partner, it was not his wife, but his minister." This became especially true after the affair of Anne Hutchinson, whose "antinomian" theology of personal revelation and aggressive challenge to the reigning religious hierarchy symbolized for many Puritans the chaos that would ensue if women were given responsibility for spiritual formation.[16]

To that end, as Marylynn Salmon has shown, Puritan colonies beefed up Patriarchal regulations to make the law more consistent in its favoring of male headship. English common law had been patriarchal, to be sure, but there were several loopholes and inconsistencies that granted some women some rights some of the time. It had been common practice, for example, to allow a wife continued authority to decide what should be done with property she brought with her into a marriage. Connecticut and Massachusetts changed this and other traditions through formal legal means so as to strengthen the husband and reduce the possibility of conflict between spouses. Fathers were given legal responsibility to mete out civil penalties to those under their charge, acting essentially as a "justice of the peace with respect to his dependents." Mothers were seen to be more emotionally indulgent and hence unfit to educate children, for education was conceived as a necessary preparation not only for material welfare but especially for salvation. The home was conceived as "a little monarchy," and it was clear who was to be king. As one Puritan put it, "our Ribs were not ordained to be our Rulers...Those shoulders aspire too high, that content not themselves with a room below the head." Womanly submission was thickly symbolized: heads were always covered, even in the home, and women and girls were segregated and silent in churches.[17]

Indeed, when one reads through some of the Puritan diaries, the little monarchy does seem to consist mainly of men. Increase Mather's diary, for example, only mentions his wife and six daughters when their illnesses interrupt his sleep or daily routine, though he spends pages and pages continually praising God for his boys, especially Cotton and Samuel. Nevertheless, by the time Cotton and Samuel came of age this tight patriarchal system was falling apart, despite increasingly desperate efforts by colonial governments to hold it together. Why?[18]

Several things help explain the shift in moral authority from father to mother. First, as Gerald Moran has shown, many men simply stopped going to church. In many cases the mother became responsible for spiritual formation by default. But why did males drop out? One speculative answer is that the rise in life expectancy and the easily available land worked together to undermine patriarchal authority in the home, church, and state. In the early days of settlement, a son's prosperity depended on the patriarch's blessing and was a necessary prerequisite for marriage. But as the patriarch lived longer and longer, thus delaying the time when the son would receive his inheritance, and given that so much good land was available just a bit west, many young men were able without much trouble at all to escape the system entirely. Thus cut off from historic forms of authority, subsequent generations of men felt less pressure to join churches because the civil benefits that had formerly been associated with membership—property and voting rights, for example—were now available without it. It may be, then, that the most fundamental shift that occurred was in how American men thought about their lives. Robert Wells has suggested that the engine driving all of the other social changes of the early modern period was the epochal shift in consciousness from a sort of peasant fatalism—a resignation to the status quo as a divinely sanctioned one that was not to be tampered with—to what he calls "rationality," the belief that the world can be known and controlled, and that individuals can improve their lives. The old Puritan religion, with its predestinarian determinism, its rigid hierarchies, and its social stratification, no longer seemed relevant to colonial men. So they left to make their fortunes; while their wives, whose shift from peasant fatalism to modern rationality was yet to occur, remained to sing the hymns and teach the children.[19]

Whatever the reason, by the end of the seventeenth century the shift from a male center to a female center for child education was well underway. And school attendance seems to have risen at about the same time. But whether it was mother or father, it remains the case that the vast majority of children acquired almost all of their formal education and pretty much everything else they learned in the home.[20]

Home Education in Colonial America

Housework itself was probably the most significant educational experience for colonial children. Though fathers were responsible early on for instruction in catechism and literacy, mothers had always been in charge of running the home. Children and servants were assembled by the mother and assigned tasks as their abilities allowed. There was much work to be done, and children did their share of it: plucking goose feathers, washing, soap and candle making, preparing and preserving food, and countless other daily chores. Girls in particular were taught the domestic arts of knitting, weaving, sewing, and so forth. To demonstrate their expertise, many girls constructed a master work called a "sampler," a rectangular piece of cloth on which were sewn various stitches the girl had mastered. Many samplers were headed with such verses as:

> When I was young and in my prime,
> You see how well I spent my time.
> And by my Sampler you may see
> What care my parents took of me.

It was not the samplers alone that proclaimed the productive potential of the colonial household. The North American colonies very quickly became so productive and successful that it disturbed the Parliament and the English Board of Trade. The colonists were supposed to be importing and consuming British goods, but instead they were largely self-sufficient. England struggled to find other ways to make money from the colonies, and their mechanisms—taxes on tea, stamps, and the like—were not well-received by the colonists.[21]

Though housework and chores took up a great deal of time, literacy was a high priority. The motivation here was largely religious. The vast majority of colonists were Protestants, committed to the doctrine of biblical inspiration. Most Protestants took it as a given that Roman Catholicism could only be successful among illiterate people who had no access to the actual Word of God. They thus greatly valued individual and collective Bible reading as the best means for maintaining true religion. But here they faced an enormous challenge, for the short history of Protestantism had shown quite clearly just how susceptible the Bible was to multiple interpretations. North American Protestantism was particularly vexed by this issue, given the large number of separatists and extremists who arrived after fleeing state church persecutions. Fearing the fissiparous and unpredictable results of their own commitment to *Sola Scriptura*, Protestant families sought to provide for their children an interpretive grid so that their Bible reading would lead to the correct conclusions. The solution was the catechism, a series of questions and answers systematically explaining Protestant theology that were to be memorized by children. Catechisms sold by the thousands, some even by the millions. The most popular by far were the *Westminster Shorter Catechism* (available to colonists by 1647) and the *New England Primer* (first available in 1690). But here too there were problems of authority, for every group had its own catechism. In 1679 Increase Mather complained that there were "no less than five hundred catechisms extant." Colonial governments might require that children be catechized and fines be imposed on parents who did not comply, but which catechism? One of the central tensions in colonial society, and indeed in American society today, pits the freedom of individual conscience against the organizational needs of the community. In one sense, the American preference for freedom of religion over imposed doctrinal orthodoxy was nurtured in these seventeenth-century families, whose fathers selected for them which catechism to use.[22]

Though colonial children were taught to read in great numbers, in the early generations there was precious little available to read beyond the Bible and the catechism. Whatever material a family did possess was read, reread, memorized, circulated, passed down to posterity, read in groups, read aloud, and read in private by candlelight during long winter evenings. One of the most popular works to be read in this way was Michael Wigglesworth's poem, *The Day of Doom* (first printed in 1662), the first in a long line of American apocalyptic bestsellers. The poem, an account of the return of Christ to judge the world, though it includes occasional glimpses of the righteous getting their just reward, lingers long over the fate of the reprobate, including "Children flagitious and Parents who did them undo by Nurture vicious." Reading works like this was its own form of education, an education that was overwhelmingly moralistic and religious, aimed at producing conversion. Colonial records are full of accounts of the emotional struggle toward conversion among the young. Samuel Sewall, for example, records in great detail in his diary the agonies of his daughter Betty, whose reading from the Scriptures and the sermons of Cotton Mather led to a breakdown, "after dinner she burst out into an amazing cry, which caused all the family to cry too...said she was afraid she should go to hell, her sins were not pardoned." Four months later her crisis had not abated, "Betty can hardly read her chapter for weeping; tells me she is afraid she is gone back, does not taste that sweetness in reading the Word which once she did." Six months later she still "weeps so that can hardly read...she tells me of the various temptations she had; as that [she] was a reprobate, loved not God's people as she should."[23]

For some, literacy was not the only formal academic training the colonial home proffered. Increase Mather recounts how he "learned to read of my mother. I learned to write of father, who also instructed me in grammar, learning, both in the Latin and the greeke Tongues." But of course the Mather family was exceptional. What the Mathers did share with almost all other early colonists was the lack of institutional schooling. Even when schools were encouraged, as in the famous 1647 "Old Deluder Satan" bill that required towns in Massachusetts with fifty or more families to have a school, the evidence suggests that most people did not comply. Historian Maris Vinovskis has compiled all the evidence to conclude that most towns chose to pay fines rather than start schools, for "the emphasis was still on having parents educate and catechize their own household members." Thus, early school laws are not evidence of the waning influence of the family but of the increasingly desperate efforts by the religious establishment to reverse their growing loss of control over families.[24]

Though the home was the basis for nearly all colonial education, that does not mean children were always taught by their own parents. One popular mode of home instruction both in the colonies and back in Europe was the "dame school," a sort of early childhood educational cooperative headed by a woman who charged a modest fee to keep neighborhood children in her kitchen, often teaching them basic literacy and numeracy. Sometimes hired servants would perform the same function (this sometimes being called a "petty school"). Call it glorified babysitting if you like, but the thing to note is that these schools were household activities.[25]

Tutoring was also popular. European theorists such as Locke and Rousseau strongly recommended the practice, especially over schools, which in their view were unhealthy and immoral. William Penn agreed, preferring to "have an ingenious person in the house to teach" his children rather than "send them to school" where they might pick up too many "vile impressions." Many other parents seem to have agreed as well, if evidence from newspaper ads is any indication. Colonial papers are full of ads such as this one, from the *New York Mercury* in 1765:

> William Elphinston
> Teaches persons of both sexes, from 12 years of age and upwards, who never wrote before, to write a good legible hand, in 7 weeks one hour per day, at home or abroad.

As historian Huey Long argues, the sheer number and frequency of such ads suggest that there must have been a brisk market for tutors making house calls. Many colonial ministers also tutored neighborhood children (usually boys) as well as their own, thus supplementing their typically meager incomes. If a boy aspired to attend one of the fledgling colonial colleges or Latin secondary schools, a tutor was all but indispensable.[26]

Another popular mode of home instruction that has been widely discussed in the historical literature, partly because it is so foreign to today's norms, was the phenomenon of sending a child to another family to live and receive an education. The practice was called by many names—putting out, binding out, placing out, fostering out—and though it was sometimes done under coercion, especially if a family were poor, a great number of colonial families put their children out to other homes for extended periods of time quite voluntarily. Why did they do this? Some historians see the practice as a throwback to chivalric days when nobility and gentry would send their children to the royal household for education. The rising merchant classes in England and elsewhere wanted to raise the status of their children and thus copied the procedure, and it trickled down from there. There may be some weight to this explanation, for it does seem to be the case that families sought in most cases to "put out up" in an effort to secure for their children access to families higher up the status chain, along with the literacy and manners such families might provide.[27]

Some historians, in contrast, stress the community-building aspects of the putting out system. Frequent child swapping no doubt made for strong bonds of reciprocity and solidarity among colonists spread out over a large geographic terrain. Others, following Edmund Morgan, see the putting out system as an example of colonial attitudes toward child rearing. We have already seen how maternal nurture was seen by many colonists as potentially dangerous since it tended toward coddling and overindulgence. Similarly, families may have balked at their ability to enforce strong discipline in their own children and turned instead to more impartial neighbors who would be less likely to spare the rod. There is some evidence for this general sentiment not only among early settlers but even among later generations. Here, for example, is William Williams, from an advice book published in 1721:

> Parents who from numerousness of their children, or low outward circumstances, are uncapable to do what is proper and necessary towards their good Education...should otherwise provide for it by disposing of

them into good and virtuous Families, where they may be well educated and fitted to serve God and their Generation.[28]

Perhaps the best way to explain the putting out phenomenon is to see it as a hybrid—one part moral and one part economic. Many parents no doubt had their children's characters in mind when they sought a fitting family in which to place them. Many colonial laws regulating apprenticeship gave very explicit instructions about the sort and duration of education host families were to provide. Masters were always required to teach children to read the Bible, and usually writing and ciphering were specified as well. But while host families were supposed to perform the "custodial and educational duties" of family life, in reality, economic pressures often trumped these concerns. At its worst, the putting out system was sheer economics—the buying and selling of child labor between adults. Formal agreements were written up, often after much haggling over fees and compensation. The higher the skill setting, the more a family would have to pay for their child to have access. Parents would also have to pay more for a younger child than for an older one, and host families sought to keep children as long as possible, for of course their usefulness would increase with their age. Given the sensitivity of such negotiations, town officials often oversaw the negotiating process, functioning as advisors, adjudicators, matchmakers.[29]

According to longstanding precedents, formal apprenticeships lasted from age seven to age twenty-one, split up into two seven-year stages. During the first stage children (usually boys) were bound to an adult and were boarded and given education in exchange for labor. At age fourteen the apprenticeship proper began, the assumption being that a boy's work was now worth more than his keep. At age twenty-one a young man came into his "majority," and was thus freed from his contractual obligations and ready to enter into a "man's estate." But in practice things often did not work out so smoothly. Many colonial children were put out well before the seventh year, sometimes even before the child could walk. Occasionally colonial courts intervened and disallowed putting out agreements because a child was deemed too young. But by far the greatest kink in the system was the unpredictability of the lifespan. Masters and boys had a knack for dying before the contract had been met, thus necessitating further negotiations. Court records also are rife with examples of fraud and failure to comply with set terms. Masters failed, not infrequently, to provide the training promised, treating the children as drudge laborers and sometimes handling them with much cruelty. Apprentices and their birth families just as frequently tried to get out of the agreement, especially as the child grew older and more useful. Most court cases involved the successful suit of a master for the return of an indentured child. Five-year-old Joseph Billington, for example, "did oft depart his said master's service" to return to his parents, but finally Billington's family was taken to court and the court immediately returned the boy and ordered his parents to be "set in the stocks" if they allowed him to return again.[30]

Over time the putting out phenomenon faded in the colonies, perhaps due in part to the decreasing economic role of the family itself as other institutions began to emerge, perhaps to the abundant land and near universality of the vocation of farming in early America. By the eighteenth century it was mostly orphans or the children of the poor who were placed in other homes, sometimes against their parents' wishes. Well into the nineteenth century the preferred way of dealing with children without parents was to place them in "orderly families." Again, the mix of familial and economic motives did not always bring out the best in people. Not infrequently, an orphan would be put on "public vendue," or auction. Local government would hand the child over to the lowest bidder who would promise, in exchange for the government money, to provide room, board, and rudimentary education to the child. Such market-driven trafficking no doubt rendered many an orphan's childhood harsh and slave-like, but it did keep such children in family settings, ensuring that they received food, shelter, basic literacy and religion, and a trade that would hopefully fit the child for useful adult occupation. And while violence against such children was common, children living with their birth parents often experienced the same thing. In fact all children, not just orphans, were seen as property by many colonists, and thus governments and neighbors were loath to interfere in cases of family violence, out of respect, not for privacy, but for property.[31]

Violence itself was for many colonists a sort of pedagogy, intentionally applied to achieve desired ends. Puritan John Robinson articulates, quite succinctly, what was for many colonists a basic assumption about the nature of children and the job of parents:

> There is in all children, though not alike, a stubbornness, and stoutness of mind arising from natural pride, which must, in the first place, be broken and beaten down; that so the foundation of their education being laid in humility and tractableness, other virtues may, in their time, be built thereupon.

This theme of breaking down the will of children has been much discussed by historians. Some, deeply disturbed, find herein the roots of much evil. One might almost say that for some historians the doctrine of original sin *is* our nation's original sin against children. Other historians see in the conservative Protestant emphasis on breaking down children's will a growing recognition that children were in fact individuals, not just property, that each had a name and a destiny for which the parent was responsible. This may be the case; but if so, it seems this dawning recognition of a child's individuality did not extend to the children of one's slaves.[32]

The issue of slavery reminds us that domestic education could be the source of tremendous conflict and controversy in the colonial America and well after. As we have seen, many Americans seemed to be committed both to the thesis that stable families were the necessary prerequisite for a stable society *and* to a mercantile understanding of the family. Though among whites these two theses gradually were resolved by downgrading the economic function of the family and moving away from the conception of children as property, the issue of slavery put these two commitments on a collision course, as Samuel Sewall noted in the late seventeenth century:

> In taking negroes out of Africa and selling of them here, that which God has joined together men do boldly rend asunder; men from their country, husbands from their wives, parents from their children.

Sewall saw more clearly than most of his contemporaries the inconsistencies of common practice. As he wrote, good, churchgoing Massachusetts citizens regularly cast aside their views concerning the sanctity of the home when there was a profit to be made. One group of Boston Negroes petitioned the Massachusetts legislature in 1774 with these words:

> Our children are...taken from us by force and sent many miles from us where we seldom or ever see them again there to be made slaves of for life which sometimes is very short by reason of being dragged from their mother's breast.

Such issues were not to go away as the colonies became sovereign states, and the consequences of the inconsistency Sewall described is of course one of the major themes of American history and current affairs. But our point here is to note that the education of black slave children, sometimes for complex trades and domestic arts, more often for hard labor, and always for docility and obedience, was home education.[33]

Inside the Great House

The plantation economy of the American South provides the most striking variety of home educations made available to different sections of the population. Consider, for example, young Eliza Lucas, daughter of a South Carolina plantation master coming of age in the late 1730s. She would typically rise at five in the morning, read for two hours, and then take a quick tour of the work getting going on the plantation fields. She would then return home for breakfast and two hours of music or French. By midmorning she would shift from being student to instructor, offering reading lessons to her sister and two slave girls. In the afternoon she would do needlework with a group of other young women and girls while one of them would read aloud, and in the evening she would write letters and read such popular authors as Milton, Richardson, and Locke. Such an education equipped Eliza to run the plantation when her father was absent, until her marriage in 1744, after which she educated her own children "on Mr. Lock's principles."[34]

Eliza's story illustrates several themes in home education in the American South. She herself received a genteel education given by a series of tutors. Tutoring was the preferred method of education in the Old South, with tutors often offering their services in "old field schools" built on the plantation for the children of the plantation owner and perhaps near neighbors. Many Southern notables, including George Washington, Thomas Jefferson, and Robert E. Lee, were educated this way. Secondly, we see here the increased opportunities for some women in the Old South, where patriarchal law was never as strong and women were often empowered and even groomed for leadership, partly because of Aristocratic pretensions and partly of necessity given the higher death rate of males in the South. Many plantation girls were given a classical education very like what was offered to their brothers throughout the eighteenth century.[35]

It is also the case that, especially during the colonial period, some plantation slaves did in fact receive a fairly liberal education, though this changed very quickly around 1831 after the Nat Turner rebellion, when teaching black children to read was outlawed throughout the South. The proscription applied to free blacks as well, who had to resort to clandestine education in the home. But in the earlier period, even slaves who were taught to read for service in the "Great House" were taught their proper place along with their ABCs. Robert Ellett, for example, describes a lesson he learned one day:

> I grew up with the young masters. I played with them, ate with them and sometimes slept with them. We were pals... One day the old master carried me in the barn and tied me up and whipped me 'cause I wouldn't call my young masters, 'masters,' He beat me till the blood run down.[36]

Finally, and perhaps most significantly, Eliza's life illustrates well the powerful influence of advice literature on American ideas of family and education. In her case, as for many others, John Locke was the guru. Locke's political influence on the founders of the United States is well known and has perhaps been overstressed. Most Americans who read Locke did not read his treatises on civil government. They read his works on education and the family. And if they didn't read Locke, they very likely read those who popularized him: Lord Chesterfield's *Letters to His Son* and John Gregory's *A Father's Legacy to his Daughters* were two of the top three bestselling books in the American colonies in 1775, and both preached the Lockean ideal of nurturing parents whose efforts can produce free and good people. It is here, as Jay Fleigelman has argued, that one finds the true seeds of the American revolution. American homes were the nurseries of republicanism long before shots were fired at Lexington. In revolutionary America, the family state remained, but it was no longer the Filmerian model of Divine right Kingship overseeing divinely sanctioned patriarchy. American families over time exchanged patriarchy for affection freely given and received by spouses and their children, and they expected the same of their government. When England failed to demonstrate reasonable affections, liberty was asserted. But as we shall see in the next chapter, American liberty brought with it tensions and contradictions so powerful that the family state ultimately proved impossible to sustain.[37]

Chapter Two

The Family Nation, 1776–1860

On any given day in 1775, you'd find Abigail Adams, soon to become the fledgling nation's second first lady, up at five in the morning to light the fire and get breakfast ready for her four children. From that point on, the day was full of domestic responsibilities for her—cooking and cleaning for the family and guests, managing the farmhands and property, overseeing assets and expenditures, hiring and governing servants, and conducting the education of her children. Schools were available in Braintree, the town the Adamses called home, but Abigail preferred her own instruction and the tutorship of her cousin John Thaxter, who boarded with the Adams family and gave instructions to the boys, and to her daughter Abigail too. After a day of tutoring, chores, and perhaps some outdoor play, the family would gather around the fire until bed. Some evenings John Quincy, age eight, would read aloud a page or two from Rollin's *Ancient History* as mother and daughter sewed, and then Abigail would take over and finish the chapter, drawing connections between ancient political events and those then taking place in the American colonies. Her husband John was so busy being a founding father that he was seldom home to father his own children, but the pair corresponded constantly, almost obsessively, about the educations and futures of their children. Every opportunity was taken to turn daily domestic activities into lessons in character and accomplishment. As one biographer put it, the parents "willed greatness on their descendants."[1]

One of those descendants, John Quincy, emerged from his evening history lessons to become the nation's fifth president. Having retired to Quincy, MA, after years of public service, he exhibited similar concern for the education of the family line. John Quincy Adams had a grandson named Henry who did not want to go to school one day. When his mother tried to force him, he began wailing, fastening himself to the staircase that led up to grandpa's study. The spat lasted for a good while, much to the embarrassment of Mrs. Adams. Then the study door opened, and John Quincy Adams came down the stairs. Henry later recalled the scene:

> Putting on his hat, he took the boy's hand without a word, and walked with him, paralyzed by awe, up the road to the town. After the first moments of consternation at this interference in a domestic dispute, the boy reflected that an old gentleman close on eighty would never trouble himself to walk near a mile on a hot summer morning over a shadeless road to take a boy to school... but the old man did not stop, and the boy... found himself seated inside the school, and obviously the centre of curious if not malevolent criticism. Not till then did the President release his hand and depart.[2]

The American Synthesis

The Adams family was exceptional in many ways, but their attitude to education was quite typical of New Englanders in the eighteenth and nineteenth centuries. The location of education—at home or in the schoolhouse, was not too significant. What mattered was that children got an education, a particular sort of education. Home tutoring, private academies, common schools supported by local taxes... all of these venues taught basically the same thing because all had the same goal of forging a common American identity from the disparate groups that made up the population. The content of this education was a synthesis of several strains of influence that on their own do not necessarily harmonize. Later generations would have to deal with the instabilities latent in the program. But before the Civil War a synthesis was forged that united even as it divided the home, school, government, marketplace, and church. To understand this synthesis is to understand much about American culture and the nostalgia many feel for the days when it all held together.

The key to the entire synthesis and the bulwark to the American theory of education was the thoroughgoing rejection of a major aspect of the Calvinist heritage of early America. Nearly as soon as Calvin's doctrine of innate depravity inherited from Adam's original sin had been formulated, it drew reactions, and by the eighteenth century the reactions had triumphed. Those who agreed with Jonathan Edwards' classic description of children as "young vipers and infinitely more hateful than vipers" had been steadily losing ground in the colonies. After the revolution, such pessimism about childhood nature was all but abandoned, despite periodic efforts to hold the line by old guard ministers. And though the ministers themselves would often blame conspiratorial atheists or Romanizers for the shift, in fact it was not the freethinkers who were transforming the American view of childhood. It was the Protestants themselves.[3]

The doctrine of original sin held, as the *New England Primer* so succinctly put it, that "In Adam's fall/we sinned all." Children were born with the curse of Adam's sin on them, and if they died before being baptized, they must spend eternity in hell. But baptism itself was not enough. Calvinism also taught the doctrine of election: only those whom God has chosen from before the foundation of the world would be saved, and this election had absolutely nothing to do with what individuals on earth do or do not do. Those predestined for salvation will be saved. Everyone else will suffer eternal torment. Some Calvinists quibbled (and still quibble) over whether God actually predestines people for hell or simply allows them to choose it, but the end result is the same: God's sovereign will is safeguarded as the prime theological principle, and human will is demoted. These doctrines received much parsing, systematizing, and embellishment from several generations of Reformed schoolmen, leading to a body of doctrine far more complex and sophisticated than I have summarized here. It is fair to say, in fact, that John Calvin sired one of the most fecund intellectual traditions the world has ever seen. But though this tradition took root early and strongly in colonial America, it did not last.[4]

There are several reasons why most American Protestants rejected the doctrine of original sin as articulated by Calvinism. Part of it was theological. Calvinism has always been a system that is difficult to believe. On an aesthetic level it has a certain beauty given its logical rigor, marshalling of scriptural evidence, and streamlined doctrine. Its unflinching willingness to hold on to harsh conclusions out of a sense of fidelity to revealed truth has always appealed to the temperamentally pertinacious. But despite all that, many people have a hard time accepting its claims on an intuitive level. One reason the Calvinist heritage has produced such a wealth of theological writing is that the system itself yearns for qualification, for relaxation. But relaxation threatens the system's inner integrity, and so every relaxation produces a reaction. The "Old Lights" reacted against the revival spirit of the 1740s. Congregationalists reacted against the Unitarianism of Channing and other nineteenth-century liberals. More recent splits within Calvinistic denominations have nearly always showcased conservatives reacting against liberalizing trends of one sort or another. In the eighteenth century such relaxations and reactions were of far greater public moment than they are today, given the far-reaching influence of Calvinistic ministers and congregations. Debates that to many contemporary readers might seem pedantic and of limited interest were front-page news then. One of the central issues in the late eighteenth century was the difficulty raised by the Calvinistic system of why humans should be held responsible for their sinfulness if God's sovereign will dictates everything. In a land brewing with unrest against established civil authority, with newspapers full of words like "freedom" and "liberty," Calvin's deterministic notions seemed out of touch. Arminians such as Moses Hemmenway argued that Adam's fall led to the loss not of man's moral nature or propensity to do good but only to "the habit or principle of righteousness disposing him to holy affections and actions." Adam just picked up a bad habit. People must be rehabituated into saving goodness until it became again "secondary nature." What was needed was not some sort of lightning strike that was utterly dependent on God's whim and could only happen to the chosen few, but a spiritual education that could extend to all people. Some Calvinists tried to counter this approach with theological arguments and proof texts, but it was hard to kick against the goads of national opinion. Many moderate Calvinists chose the route of relaxation. Perhaps the most popular treatise written by a Calvinist minister in the early 1800s, Horace Bushnell's *Christian Nurture*, became so beloved by the laity (and hated by conservative ministers like Charles Hodge who considered it Pelagianism) because it provided a clear description of the power parents have to cultivate Christian character in their kids. For Bushnell it was not the revival conversion or even the primordial election of God that produced good Christians. It was good parenting.[5]

Theological reasons for rejecting doctrinaire Calvinism were nearly always bound together with cultural concerns. America was a much more comfortable place to live in the late eighteenth century than it had been in earlier times, and most folks were less concerned with doctrinal purity than with getting along with neighbors and making their colony, and later their state, economically successful. What was needed was a Christianity that would bring people together and provide religious sanction for their efforts toward economic progress and social stability. Many historians have noted a general softening of tone at all levels of American society throughout this century. In the 1760 version of the *New England Primer*, for example, the ominous "Duty of Children to Parents," which threatened disobedient children with damnation, was replaced with the "Cradle Hymn" of Isaac Watts, which opens with the maudlin couplet, "Hush my dear, lie still and slumber / Holy angels guard thy bed." Isaac Watts' influence on

American Protestantism extended well beyond the best-selling primer. His Christianized psalms set to music proved tremendously popular and transformed American worship. Prior to Watts, many American congregations had sung only from the Psalter, associating hymns not directly found in the scriptures with Catholic man-made traditions. But Watts himself found many of the Psalms in their original "almost opposite to the spirit of the gospel: many of them are foreign to the state of the New Testament." So he changed them, eliminating David's darker thoughts and appeals to ethnic violence, substituting instead Christian consolation and comfort. His collection went through twenty-eight printings in the New World between 1770 and 1783 and paved the way for his original compositions. Having quickly accustomed themselves to singing heavily paraphrased psalms, congregations went one step farther and allowed the inclusion of other texts that preached true religion. The hymns of Isaac Watts were "apparently the first nonscriptural verses to be sung in eighteenth-century congregations. The closed world of the revealed word had at last been opened to a new human voice and text." It is astonishing that such a profound shift occurred so easily.[6]

And so widely. Historian Nathan Hatch has described in great detail how populist preference for exuberant worship, lowbrow preaching, and a theological stress on human freedom triumphed over old guard Calvinism's severe restraint and fussy doctrine in the decades after the American Revolution. Whether or not one ought to call this far-reaching transformation a "Second Great Awakening," it is clear that American Christianity was changed and strengthened during these decades. Before the Awakening there were about 300 Methodists in the country. When the indefatigable Francis Asbury died in 1816, there were 200,000. By 1812 there were also 200,000 Baptists, and both groups continued to grow exponentially: by 1850 there were over a million Baptists in the United States. Indeed, according to some historians, it was only after the American Revolution that the American masses could really be said to have become largely Christianized. And the Christianity that was embraced by millions stressed free will over predestination, social betterment through moral reform, and the divine blessing of the United States of America. Why was this sort of Christianity embraced so willingly? Partly, as Hatch notes, because it harmonized nicely with democratic sensibilities. Partly, as Ann Braude has noted, because "in America, women go to church," and these orientations seemed to click with many American women. A European visitor to America, Frances Trollope, was shocked by a visit to a revival camp meeting, writing in 1832 that she "never saw, or read of any country, where religion has so strong a hold upon the women, or a slighter hold upon the men."[7]

But democratic sensibilities and the feminization of Christianity only go so far in explaining the far-reaching impact of revivalist, populist religion. The greatest gift this more generous, less stringent form of Christianity offered to aspiring democrats and mothers was a way to reconcile their desires to have their children be both Christian and successful, poor in spirit yet rich in earthly things. The earlier, harsher stress on dramatic conversion, passive submission to divine providence, and unquestioned obedience to authority simply did not provide American families with an effective way of adapting their children for success in the increasingly urban and industrial society that was in the making. Nearly all Awakening ministers stressed "the necessity for greater familial education," emphasizing the nurturing role parents could play in stimulating positive moral traits in their children. For a nation of farmers experiencing the straining of traditional patterns of living due to the exhaustion of land, population increase, and the growth of trade, a Christianity that celebrated human free will, sidestepped formal clerical hierarchies, spoke with a bucolic accent, and provided a network of friendships and institutions to help them navigate the modern world proved very attractive. Evangelical entrepreneurs figured out very quickly that messages of comfort and hope were received with far greater enthusiasm than the older fire-and-brimstone fare. Mothers wanted a vision of childhood that spoke not of vipers and hell but one that, in the words of an 1833 piece in *Ladies' Magazine*, "speaks to us of heaven; which tells us of those pure angelic beings which surround the throne of God, untouched by sin, untainted by the breath of corruption."[8]

The shift from belief in innate depravity to childhood innocence was connected to many other shifts in family life in the early nineteenth century. Historian Charles Sellers has described well the gradual replacement in the American interior of an agrarian, land-based economy of subsistence with a mercantile, profit-driven, capitalist order. Cities emerged to consolidate economic output. Canals and rails were constructed to transport goods. Capital, goods, and labor thus became increasingly mobile, and the entire process was abetted by pro-market legislations and court decisions. This "market revolution" had a profound impact on American families. Families became more isolated from one another and increasingly dependent on the nuclear family itself to provide emotional nurture and companionship historically found in the wider community. Rising prosperity seems to have correlated with an increasing standard of privacy: middle-class children began to have their own beds and spaces; homes became larger with distinct rooms for distinct activities; household activities became increasingly separated from economic production; birthrates declined as children became more an economic liability than an asset and parents adopted more labor-intensive child-rearing techniques.[9]

Such changes didn't happen overnight. They occurred in some places earlier than others and tended to affect the town-dwelling middle class more obviously than rural populations. People who write history books tend to be most interested in dynamic transformations, especially in the shift from premodern to modern life, so their accounts often overstress the cutting edge and ignore the less interesting story of continuity or stasis. Many families continued to make their own clothing, hosiery, linens, and candles. Many kept stocking their larder, rendering their fat, and visiting their neighbors. Many did not move west or into cities. But even among "those who stayed behind" and kept farming, the trend was in the direction of specialized, marketable crops and animals rather than self-sufficiency. Capitalism was transforming not only the townies but the way farming was done as well. In fact, some historians have argued that American-style capitalism was born more on the farm than in the town, as farmers shifted early and eagerly from producing the full range of food their families needed to growing the crops or raising the animals they thought would make them the most money. Farmers still maintained traditional cultural practices and voiced old-fashioned opinions, but their own specialized, efficient cultivation of crops for market both pioneered entrepreneurial methods and provided cheap food for other entrepreneurs so they could develop businesses in other domains.[10]

Specialization, or to us another term, segregation, increasingly characterized many noneconomic aspects of life in postrevolutionary America as well. The word "segregation" carries pernicious connotations for us today given its association with Jim Crow. But for many nineteenth-century Americans it was an addictive habit given its obvious advantages in agriculture and other industries. It was also a great tool to help reconcile, if only for a time, opposing tendencies in their society. Most profoundly, the Lockean instinct that formed the basis of the American political experiment, the assumption that all are created equal, simply did not square with social reality, for women and slaves were clearly not politically free or equal. The solution was to segregate, to claim that people are free to be not whatever they want to be but to be who they naturally are. Just as some climates fit naturally with wheat cultivation while others favor cotton, so different groups of people have different political roles. Politically for white males the Lockean notion of freedom as an unbounded possibility became the norm, but for the private sphere an older, more Aristotelian notion of the given place in the natural order for women and slaves was maintained. In the Colonial period, slavery had been practiced mostly as a "socially necessary convenience." Work had to be done, and slave labor had long been a good way to get work done. But by the late eighteenth century slavery had been rationalized, theorized. It had to be. If the new nation is premised on freedom, then why does it countenance slavery? Because, many reasoned, black people are biologically inferior to whites and should therefore specialize in slave labor. The existence of free blacks refuted this view, so they were repressed. It was only logical. If all men were created equal and if slavery were legitimate, then blacks must not be men, and of course neither were women.[11]

But if women were not free, they were the pivotal character in keeping society itself sane. Women, too, had a specialized role to play in the new nation. As marketplace values increasingly characterized the public world of politics and work, many Americans feared that the moral character necessary for republican government would not survive. It was left to mothers to be the guardians of public virtue, what Daniel Webster called the "effective teachers of the human race." Women, though not the equal of men, would have to be educated themselves so as to train the men. "The mother forms the character of the future man," asserted Catharine Beecher, "if this be so, as none will deny, then to American women, more than to any others on earth, is committed the exalted privilege of extending over the world those blessed influences, which are to renovate degraded man and 'clothe all climes with beauty.'" While not equality by any means, this perspective did turn domesticity into an intellectually stimulating and spiritually significant function. It provided a rationale for female education that even conservatives could support and used the home to purify public life even as it kept women and children tucked safely inside. This is the family that the French visitor Alexis de Tocqueville found in America in the 1830s, not the "little commonwealth" of a patriarch presiding over his realm, but the "haven in a heartless world" of nineteenth-century separate spheres dualism between the male cutthroat domain of work and politics and the female home of morality and tenderness. Despite the fact that American women were circumscribed "within the narrow circle of domestic interests and duties," Tocqueville found that they rather liked it. The Frenchman was adamant that America's growing strength and prosperity ought be attributed to the "superiority of their women."[12]

Fathers did not exactly check out, but increasingly they delegated (or simply defaulted) moral authority to their wives. The mother became the household's soul and body, if not its spiritual head. She was "God's own police," as one newspaper put it, her home a moral haven guarded and protected by her vigilant eye and faithful prayers. And her beneficent influence would ripple out from the home to the rest of society. "Women are... the civilizers of mankind," wrote Ralph Waldo Emerson, "What is civilization? I answer, the power of good women." While many historians understandably have seen in such shifts a "feminization of American culture," some note that the softer tone of family, church, and public life was not necessarily tied to gender. Daniel Smith and Ted Ownby, for example, have both shown for very different social groups that many men participated deeply in their children's moral lives and did so in a very nurturing manner. For some historians, the shift from stern punishment to soft discipline is less an example of feminization than of a broader "humanitarian revolution" affecting all sectors of modern society characterized by a growing sympathy for human pain and aversion to cruelty.[13]

Gendered or not, humanitarianism's beachhead was undoubtedly the home. Domesticity became increasingly important after the Revolution throughout American society. Idealized family times in the mornings and evenings where father or mother read from the Bible and led the family in prayers were encouraged. Novel institutions like the family vacation and family-oriented celebrations like the birthday party, Christmas, and Thanksgiving were invented or took on their modern form, shifting from community carnival that often mocked the social order to private "sentimental occasions" that reinforced it. Whether or not most families actually looked like the image that was constantly recreated in print, such was certainly the ideal to which most aspired then and many aspire still. American mothers hungered for advice about how to carry off this successful domestic life that seemed to be the key to both personal happiness and social stability. And they got what they wanted.[14]

The most popular books of the Revolutionary and Antebellum periods were often the ones attempting harmonizations of Protestant Christianity with the more modern sentiments of Locke or Rousseau concerning the nature of children. Parson Weems' *Life of George Washington* was so well received partly for the memorable moral anecdotes it fabricated for the young George. But the exchanges between George and his father Augustine also portray the elder Washington as the quintessential Lockean dad, inculcating moral truth not by force or long-winded sermons but by example and object lessons adapted to the child's outlook and sentiments. The phenomenally popular novels of Samuel Richardson likewise present a parental ideal that is far removed from the older patriarchal style but still profoundly moralistic and Christian. And dozens of authors tried their hand at more direct advice books aimed at a growing audience hungry for wisdom about how to teach their children to be both Godly and successful. Most of the literature popularizing the "new humanitarian sensibility" headquartered in the home was written by what Nathaniel Hawthorne famously called "a d——d mob of scribbling women." Antebellum women could choose from some thirty periodicals written and edited by American women on family themes, and some of the authors became true national superstars: Lydia Child, Lydia Sigourney, Catherine Maria Sedgwick, Sarah Josepha Hale, Catharine Beecher, Emma Willard, and Almira Phelps are standouts.[15]

Domestic advice literature written by and for women dwelt mostly on instructing parents in how to educate their children. The authors held exalted views about the potential of the family to redeem individual souls at home and thus produce collectively a virtuous society. Such sentiments were not new. Many a Puritan preacher had argued, in the words of Samuel Stoddard, that "if we have rude families, we shall have a rude country." As the Puritan ministers were fighting against a stream of prosperity and secularization in their day, so the domestic educators of the 1830s and 1840s were trying to recover from a similar loss, and for similar reasons. Cotton Mather back in the early 1700s had explained the decline of conservative religion as a product of economic success: "Religion brought forth Prosperity, and the daughter destroyed the mother." Antebellum New England was similarly prosperous, leading to a preference for Unitarianism and "fashionable religion" among intellectuals and the rising middle classes. The leading domestic educators were Congregationalists eager to recover their losses and return the country to a more evangelical path. They were convinced that if they could teach mothers to teach their children, the country could be kept on the straight and narrow.[16]

What did they teach parents to do? Historian Barbara Beatty's study of dozens of the domestic education books of the antebellum period found several consistent themes present. The books are overwhelmingly moralistic, are addressed mainly to mothers, and urge an informal teaching style that emphasizes objects and experience over direct sermonizing and passive listening. A good illustration of these very points is the nascent "children's literature" being produced for

American children's direct consumption at this same time. Evangelical authors in years past had often railed against fairy tales and other fiction because they corrupted children by leading them "away from single-minded devotion to a God who had created only one world and one truth with fixed categories of being and identity." As nineteenth-century evangelical educator Lyman Cobb had put it, "dialogue between wolves and sheep, cats and mice...is as destructive of truth and morality as it is contrary to the principles of nature and morality." But the stuff was so popular that many evangelicals changed tactics and decided to use fanciful media to impart a pious message. What we now call children's literature was born of this congress of the traditional fairy tale with the moralism of antebellum evangelicals. The story provided a pleasant medium for the pious message, and this is exactly what mothers were asked to do in their daily lives. Taking inspiration from the famous Swiss pedagogue Johan Pestalozzi, domestic educators urged parents, mothers especially, to make a school of the home and fireside. Every moment, every experience might become an opportunity for inculcating morality and intelligence into one's children, not through harangues and lectures but through subtle suasion and example.[17]

So in one sense the domestic education movement of antebellum America was a conservative reaction against the materialism and individualism of an emerging industrial order and a political life of rampant corruption and coarseness. It wanted to keep women at home, children good, and the nation well-mannered and pious. But in raising the status of domesticity it was also quite progressive in many ways. If domestic women were to achieve their task, they would need to be well informed about social problems, health and nutrition, child psychology, scientific advances relative to housekeeping, and much more. Authors such as Catharine Beecher, whose 1841 *Treatise on Domestic Economy* provided mothers with the first comprehensive American volume on all of these themes, met this need for authoritative advice. But ironically, the success of such books produced a class of very public women with substantial independent income and an influence extending far beyond their own homes. The precepts of domestic educators may have taught women to stay at home, but their own lives suggested a more expansive purview. And as a further irony, it was the profiteering male-dominated publishing world that printed and marketed such books to women. Domestic education manuals tried to help women save the country from the values of the marketplace, but it was the marketplace itself that brought the message to its audience.[18]

Home and School in Antebellum America

The American synthesis we have just described was full of latent tensions that would eventually prove its undoing. Antebellum Americans thought that they could be both Christian and capitalist, that they could serve God and mammon. They thought they could hold on to historic divisions between male and female, white and black, maintaining Lockean boundlessness for white men while women and blacks were left with a bounded liberty premised on biological subordination. And they thought, or at least hoped, that the institutions they were creating would be enough to hold it all together. By the 1840s it was becoming clear to many middle-class Anglo-Americans that something more than strong families would be required. Between 1840 and 1850 the immigrant population increased by 240 percent. Many of these immigrants were Irish Catholics and other groups whose home cultures were very different from that idealized by the American synthesis. So Americans created public schools.[19]

We must note very strongly here that the people who created the nation's public schools were not conspiratorial freethinkers out to destroy the Christian nation. They are sometimes portrayed as such by critics of public schools today who read current controversies back into the past. Horace Mann in particular is often singled out as the chief villain who wrenched schooling away from families and churches and put it in the hands of government, creating a socialist system of education that has all but destroyed our nation. It is true that Horace Mann was a Unitarian for whom education was something of a surrogate religion given his loss of faith in Christian teaching, and some of his critics did accuse him of creating Unitarian parochial schools at public expense. But Mann, as important as he was in the history of public schooling, is really an outlier here. The overwhelming majority of reformers and advocates for public education in every state of the Union were evangelical Protestants, many of them ministers. In historian Carl Kaestle's words, "they were characteristically Anglo-American in background, Protestant in religion, and drawn from the middling ranks of American society." The founders of the public schools *were* the Christians, the same sort of people who were writing and reading domestic education literature. Though today many pro-family groups are pitted against pro-school groups, in antebellum America and long after, the advocates for home and hearth and the advocates for strong public schools were the same people. Protestant missionaries spread the evangelical gospel throughout the American West, creating in the process what historian Timothy Smith has called "a new religious synthesis" which would give "members of the diverse sects a common Faith." The common faith was preached uniformly in public schools, Sunday schools, the pulpit, and a host of formal and informal associations that together made up what historian Charles Foster called "the Evangelical united front." Protestant reformers knew that they could not formally establish their brand of Christianity as the official religion of the nation. They knew as well that "the survival of the republic depended upon the virtue of its citizens." They tried through the schools and other institutions to encourage voluntary adoption of the American synthesis by all.[20]

Why did Americans choose public education? There are several reasons. First, it is important to note that many Americans had been sending their children to school for a long time before state laws were passed granting free schooling to all paid for by universal taxation. Protestants had long seen schools as a necessary accessory to true religion, for in their minds Catholicism was associated with ignorance and illiteracy. To be well educated was to be a Protestant. At the same time, schooling was seen by many to be a wise investment in a child's financial future. This seems to have been especially the case for mothers. Fathers sometimes were only interested in getting as much work out of their boys as possible, focusing more on the short-term economic situation of the household than the long-term future of individual children. Mothers, more farsighted, would lobby for formal schooling, especially for second and successive sons who would not inherit the family farm. For these boys, academic training was often given in lieu of a tract of land. Some farmers, frustrated by the hard work and constant uncertainty of the job, dreamed of something more secure for their offspring. Eason Eaby, a farmer in Washington Territory, explains it well, "This season the wheat is badly blighted, potatoes being eaten up by the bugs and the blight among them too ... I am greatly concerned about the children's future—and would like to go somewhere so that they could secure a good education." As Kansas farmer Frank Klingberg observed, "the hardships of farm life caused parents to plot professional life for their brightest children."[21]

But perhaps the biggest reason Americans had long been patrons of schools was the age-old problem faced by mothers needing some time to do things other than take care of the kids. Schools for many were a place to dump the kids off for a while and hopefully get their minds improved in the process. Lucia Downing, a teacher in Vermont in the 1820s at a local district school, noted, "parents were glad to be relieved of the care of their offspring, and no one ever suggested a shortening of the hours." And Lucy Larcom, a student in Massachusetts in the 1820s recalled how she "began to go to school when I was about two years old, as other children about us did. The mothers of those large families had to resort to some means of keeping their little ones out of mischief, while they attended their domestic duties." So ready and willing were many parents to be rid of their children, in fact, that one of the loudest complaints of school reformers throughout this period concerned the "failure of parents" to shoulder their own portion of responsibility to educate their kids. The "neglect of parents," editorialized the *Brattleboro Eagle* in 1846, was the "greatest obstacle" to success in school.[22]

Another reason for the easy adoption of schooling was the level of trust between school and home. In 1830, of the nation's inhabitants, 91 percent lived in rural areas or towns populated with less than 2,500 persons. As late as 1860 the figure was still 80 percent. Throughout the nineteenth century most Americans lived in small towns or rural areas with high levels of racial, class, and religious homogeneity, where everyone knew everyone else. Teachers were often relatives of local citizens, appointed by the town, boarding in the homes of townspeople. The schoolhouse was placed in a central location, "as near the home as possible" with the intent being "to carry the home, as it were, into the school," as a Swedish visitor to the United States noted in 1840. The main thing most parents wanted from these schools was good discipline. When strangers were hired by local families, they were often interrogated sharply to ensure that the teacher would be able and willing to provide firm discipline. As one parent advised a prospective teacher in New York, "Cuff him, thrash 'em, any way to larn'em, but whatever you do, don't let 'em thrash you." One Georgia man reminisced about his schooling in the 1840s, "I never had to be punished by a teacher; not because of my 'goodness' but our parents suggested that if our teacher had to punish us we would get another punishment at home."[23]

When proposals were presented to turn schools like these into free institutions and to improve their quality, most parents were delighted, at least in the north and west. Many southern states did not share this historic attachment to local schools and were far less excited about increased taxation. Public schools did not become standard in the South until after the Civil War when the North required that each state must have a public school law for readmission into the Union. And in border states like Indiana, where Southerners and

Northerners shared the territory, there was sometimes acrimonious debate about proposals to fund schools through taxation rather than private tuition. It is important to note here that debate was not so much over schooling itself but over taxes. Reformers advocating tax-supported schools marshaled several arguments to convince their publics to vote to tax themselves, but perhaps the most successful and important one also helps explain why eventually even those who were skeptical of free schools bought into the program. The argument was that tax-supported schools would not only be better but *cheaper* than schools funded by private tuition. They would be so because they would be staffed not by family-wage earning males but by women who would bring a more tender, domestic spirit into the schoolroom, and do so for a much lower fee.[24]

It was the female teacher especially who succeeded in making the school "the great auxiliary of the fireside" in the words of domestic education advocate Samuel Goodrich. Female teachers brought a maternal presence to the classroom, turning the school into an extension of the home. These teachers were typically young, unmarried, idealistic "soldiers of light and love" keen on spreading the evangelical gospel of patriotism, Protestantism, and social uplift to their charges. School teaching was a missionary endeavor, especially on the western frontier and in the South. "I go West to do the Will of my Heavenly Father," wrote Michigan-bound teacher Flora Davis Winslow.[25]

It was a lonely calling, and many teachers did not last long, but they did have resources at their disposal. They had their strong faith and the good will of backers in the eastern states. But perhaps most important was the wealth of textbooks that made it easier to reinforce the American synthesis in their classrooms. Textbook publishers discovered quickly that pious, patriotic fare could generate astonishing profits, so they delivered. Historian Ruth Miller Elson, in her exhaustive study of these nineteenth-century textbooks, concluded that all of them consistently upheld the following values, "love of country, love of God, duty to parents, the necessity to develop habits of thrift, honesty, and hard work in order to accumulate property, the certainty of progress, the perfection of the United States." Here, as a typical example, is John Bonner's account of the Founding fathers in his *Child's History of the United States* (1855):

> If you seek to know why your countrymen have outstripped all the nations of the earth...the reason is easily found. The founders of this nation were honest, true men. They were sincere in all they said,

upright in all their acts. They feared God and obeyed the laws... Above all they insisted, from the very first, on being free themselves, and securing freedom for you, their children. If you follow the example they set, and love truth, honor, religion, and freedom as deeply... the time is not far distant when this country will far excel other countries in power, wealth, numbers, intelligence, and every good thing.[26]

Americans chose public schools, then, because they seemed to fit well into the American synthesis. The mother would train up the child at home in godliness. The Kindergarten and other infant schools would supplement the home, turning play into education in a "new environment appropriate to young children." The schoolteacher and her pious textbooks would give all children a common American and Protestant orientation. The Sunday school would serve the important function of providing particularistic denominational training, key for enabling Americans "to reject denominational schooling" the rest of the week. Thus children growing up in pious Protestant homes would have the moral culture of the home reinforced in the school, and children growing up in homes that were not Protestant or pious would attend schools as a corrective. For both groups the schools would be free of charge. To most Protestants it seemed like a great deal.[27]

But to Catholics and some other groups not comfortable merging into the American synthesis, it was not a great deal at all. Many of these people were paying taxes to support schools that were systematically undermining their home cultures. In many parts of the country, especially densely populated urban centers with high levels of non-Protestant immigration, debate and even rioting occurred over such issues as whether Catholic children should be forced to read from the King James Bible or to recite the Ten Commandments in their Protestant form. Though many efforts at compromise were made between Catholic leaders and government officials, Protestant public opinion in the nineteenth century was so opposed to granting even the most basic concessions to Catholics that the Church's bishops ultimately chose to create their own alternative school system rather than suffer under what they rightly took to be Protestant government schools. Though they complained bitterly about having to pay taxes to support schools that were hostile to their faith, millions of Catholics from the 1850s on paid their taxes and then paid again to send their children to parochial schools.[28]

Much more could be said about the conflicts regarding Catholics and other marginal cultural groups for whom the public schools of nineteenth-century America were anything but common. But the point to note here in a book on education in the home is that no Catholics, German Lutherans, Orthodox Jews, or any other sectarian group that found itself outside of the American synthesis rejected schooling as such. The specifics of what went on in schools were debated, but no one was questioning the idea of school itself. By the nineteenth century, schooling was so universally accepted as an institution that its presence was hardly ever questioned and its adoption simply assumed. Even utopian communes such as the Quaker and Oneida communities in New York, experiment as they might with celibacy or free love, taught the children of their compounds in traditional schools. While in England there had been and continued to be a lively debate over the relative merits of private tutoring at home versus institutional schooling, in America the issue was seldom raised, especially after the 1840s. There was a bit of debate as to whether schools were good for girls, but here too public opinion largely accepted the notion of schooling and even of coeducation without too much fanfare or controversy. The result was that whereas only about half as many women as men were literate in 1790, by 1870 girls had surpassed boys in literacy and academic achievement.[29]

Yet despite the formal consensus regarding schools, many nineteenth-century Americans continued to practice more family-based education. Many of those living in sparsely populated areas did so only because schools were not available or convenient for them. But even in cities and towns, many among the middle and upper classes continued to patronize private schools and to employ private tutors. They seemed to do so both on moral and intellectual grounds. Education for many in antebellum America was primarily about individual self-improvement. Diaries record intense programs of private reading, language acquisition, and scientific training for many young Americans of means. In fact, as Margaret Nash has shown, it was the desire for improved scientific training (and the expensive apparatus required to get it) that convinced many wealthy Americans to break with their historic preference for private tutors and opt instead for academies. But it was not only intellectual attainment that animated the withdrawal from the common school. Lydia Sigourney articulated what was for many their chief worry about schooling:

> Why expose [the child] to the influence of evil example?...Why yield it to the excitement of promiscuous association, when it has a parent's house, where its innocence may be shielded, and its intellect aided to expand? Does not a mother's tutoring for two or three hours a day give a child more time than a teacher at school?[30]

Sigourney here is restating the classic argument of John Locke that schools tend to introduce vice and corruption into a child's life too soon and thus a tutorship at home is the superior method of instruction despite the naiveté that often results. This argument, with its wisps of aristocracy, never caught on well in the United States as public policy, but it did inform the practices of some families. Ironically, some of the very people pushing so strongly for common schools that would raise the masses up to the level of the middle-class Protestant consensus were tutoring their own children at home out of a fear that these very masses would corrupt their own kids. One such individual was Horace Mann himself, whose wife Mary taught their three children at home even as he stumped the country preaching the common school. Mann's biographer Jonathan Messerli captures the irony well:

> From a hundred platforms, Mann had lectured that the need for better schools was predicated upon the assumption that parents could no longer be entrusted to perform their traditional roles in moral training and that a more systematic approach within the public school was necessary. Now as a father, he fell back on the educational responsibilities of the family, hoping to make the fireside achieve for his own son what he wanted the schools to accomplish for others.[31]

In the antebellum South, as noted in the previous chapter, tutoring continued to be the preferred means of education for the elite. But such was the case in other parts of the country as well. Selwyn Troen's detailed study of the St. Louis school system, for example, reveals just how persistent older patterns could be. Long after St. Louis had public schools, many middle- and upper class children continued to be taught at home by tutors, several of whom had been brought from Europe for this very function. The specifics of study might change over the decades (Southern girls got less Latin and more embroidery lessons in the nineteenth century, for example), but there is a remarkable and little-noted continuity of form from one generation to the next among the quiet American aristocracy.[32]

There were always a few people who chose to teach their children at home out of dissatisfaction with the schools available. One such person was Nancy Edison, who, frustrated with the rigid discipline and insensitivity of the teacher at the school her son Thomas attended in Port Huron, MI, decided she could do better herself. Mrs. Edison started her son on a course of great literature—by age twelve he had read Shakespeare, Dickens, Gibbon's *Decline and Fall*, Hume's *History of England*. Noting an interest in scientific topics in her son, she procured for him Parker's *School of Natural Philosophy*, which taught how to perform basic chemistry experiments. Thomas Edison soon became a compulsive chemist, spending all his money on chemicals and apparatus to stock the laboratory he built in the family's cellar. He later recalled, "My mother was the making of me. She understood me; she let me follow my bent."[33]

For most Americans, however, home-based tutoring was something only done in a pinch. The strong preference was for schools. This is especially evident when considering the westward migration of frontier families. Though you wouldn't know it watching Hollywood movies, the West wasn't won by gun slinging cowboys and lawless drifters. "The true settlers of the West," writes historian Elliott West, "came as families. It was because of his family that the male pioneer was willing to build a town and make a farm or ranch." Pioneer families tended to follow a predictable pattern of education for their children. First of all, there was of course the nonstop work necessary to keep everyone warm and fed. Secondly, there were the long evenings and winter months to endure without the amusements that technological advance would later provide. "Our winter evenings were largely spent by the fireside, mother sitting with her sewing and mending, and the boys seated on a brick hearth fashioning with their jackknives cunning little cedar boxes, listening as father read to us" reminisced Wisconsin settler Charles Weller. When it came to formal education, families would typically begin with home education but then work toward creating a school as soon as settlement was dense enough to support it. The experience of prohibition and women's suffrage activist Frances Willard illustrates the seamless transition from home to school.[34]

Willard lived on a farm outside of Janesville, Wisconsin, from her seventh to her nineteenth year. For the first two years of settlement she had "no special recollection of books," since her parents were busy setting up the farm. In 1848, When Frances was nine, and with no school available in their area, the Willards built a pine table, put it in the parlor, and invited the two Inman girls who lived a mile away to attend school with Frances and her sister. The family hired a local young woman, who had had some schooling in the East before moving out West with her family, to come and teach the foursome in the summers. For two years this school met in the Willard home, growing a bit in its second year. The next year Mr. Willard and Mr. Inman built a "plain and inviting" little building for their growing group of students. For a young girl who had never spent much time outside of her home, this modest edifice about a mile's walk away "was a wonder in our eyes, a temple of learning, a telescope through which we were to take our first real peep at the world outside of home." The parents procured for this school a Yale graduate who had tutored at Oberlin for a time. Frances learned at this school rudimentary mathematics and literature as well as some things she hadn't anticipated. One day a classmate volunteered information on intimate topics from which Frances' mother had always shied away when they came up. "It was a rude awakening, one that comes to many a dear little innocent of not half my years, and is morally certain to come if a child goes to school at all." But for all this, young Frances regarded the move from her home to the school by the river, "the great event of my life...to go outside my own home and be 'thrown upon my own resources.'"[35]

Willard's experiences were quite common in the Western territories. Frontier mothers were very often the first and sometimes the only teachers of their children. In some locations young women served as "circuit-riding tutors" to frontier families, going from homestead to homestead offering itinerant instruction to rural children. Such circumstances were gradually eliminated as settlements grew denser, but even today isolated families receive house calls from public school teachers, often by plane, in the remoter regions of northern and central Alaska.[36]

Other People's Children

We have seen in this chapter how the public school became one institution among many created by American Protestants to ensure that the American republic stayed strong and virtuous. Though some blue-blooded Americans didn't patronize it themselves, they thought it just the thing for other people's children, especially the children of immigrants. But what of children who were even further removed from the American mainstream? Another institution with aims similar to the public school was the orphanage, which gradually replaced the family-based indenture system of earlier years. Why? As noted earlier, the economic exchange of children by individuals often led to child abuse. This uncomfortable reality became increasingly intolerable given the humanitarian sentiments of the American middle class and its romantic view of childhood as a stage of holy innocence and malleability. Between 1820 and 1860 in the United States, 150 private orphanages were founded. Such institutions created problems of their own that would lead eventually to another reversal of public policy and sentiment, but the popularity of orphanages in the early nineteenth century serves as a good example of the widespread optimism during this period that institutions could solve besetting social problems. It was hoped that children without parents could, through the institution's modern methods and technique, be molded into law abiding, religious, and hardworking citizens.[37]

But even as Americans were moving toward an institutional solution to the problem of orphans, some were moving in the opposite direction regarding the education of native populations. American Protestants had a very difficult time for centuries converting natives to "civilized" ways. Day schools run by missionaries had produced very few converts either to the faith of the sponsoring group or the ways of the white man. In an attempt to reduce the relapse of children "into their former moral and mental stupor" when they returned to the village at night, some missionary organizations experimented with the idea of "adopting" native children into the families of their missionaries. In Minnesota, for example, Protestant missionaries brought about fifty Dakota children into their own homes between 1835 and 1862. The children were deliberately isolated from their home culture, scrubbed down, given western clothes, haircuts, and names. They attended schools run by the mission during the days and then received further education for civilization after school in the homes of the missionaries. Girls learned sewing, spinning, cooking, and cleaning, while boys learned farm work and outdoor chores. Extensive religious instruction pervaded the entire experience. Perhaps the most effective tutors of these native children were the missionaries' own kids, whose policing, English-speaking, and lifestyle-modeling were more effective than any number of sermons or formal lessons from adults. Missionary kids Eliza and Mary Huggins, for example, took personal responsibility for the two native girls living with their family: "we took them into our prayers. Although there were only four of us, we had little prayer meetings every Sunday afternoon, and we felt great spiritual exaltation." Shortly thereafter the two native girls joined the mission church, and they were not alone. Children who boarded with missionaries were far and away the most likely converts to Protestantism. Many of them went on to marry Euro-Americans and to live as "civilized" Christians. But the racism of nineteenth-century society made life difficult even for the most obeisant convert. To many white Americans, a civilized Indian was still an Indian. For some native converts, the ridicule and marginalization they faced upon full entry into white society drove them back to the tribal life they had left behind, much to the horror of the missionary families who had spent so much time and effort on them.[38]

If home education was used by whites to assimilate natives, it was used by blacks to resist oppression. Antebellum slave states passed increasingly draconian antiliteracy legislation for their black populations, both free and slave. Training in literacy thus had to go underground. Some free blacks sent their children to relatives in the north to learn to read and write. Others engaged in clandestine activities centered in the home. Susie King Taylor of Savannah, GA, where state and city law forbade teaching literacy to any person of color, camouflaged her school books and snuck into the home of a free black woman each morning, dodging police and other white people. She joined with twenty-four other children who learned to read in the kitchen of Susan Woodhouse. After lessons were completed, the children would slowly disperse one at a time, sneaking back to their own homes. Sometimes masters' wives conspired with slaves to teach them literacy over the objections of their husbands, often out of concern for their slaves' salvation. Slaves sometimes bartered with white children—trading food and money for letters and words. G. W. Offley, for example, provided food for a white child whose father had gambled away the family's money in exchange for writing lessons. James Fisher gave an old man whiskey money in a similar exchange. Slave boy Richard Parker collected old nails that he exchanged for marbles that he gave away to white children in exchange for help with spelling lessons in the book he carried under his hat at all times. Alice Green learned to read by asking the white children what they had learned at school every day once they returned home. Allen Allensworth did the same by getting the master's son to "play school" with him in the afternoons. One slave's learning would quickly be shared with others. Contemporary excavations of slave quarters have found, along with pottery and furnishings, the remains of graphite pencils and slate tablets. Some of the tablets still have recognizable words and numbers on them. Much of this slave-to-slave transfer of learning would take place on Sundays when the master and family were away at church. One slave who had been taught to read in secret by the master's son recalled, "when my master's family were all gone away on the Sabbath, I used to go into the house and get down the great Bible, and lie down in the piazza, and read, taking care, however, to put it back before they returned." Encounters like this helped African Americans cultivate a Christian faith premised not on obedience to authority but on the promise of freedom—freedom like that of the Israelites from Egyptian bondage, freedom in Christ in whom "there is no slave or free."[39]

But such lessons only heightened the jarring inconsistencies and ironies of the American synthesis. In individual homes, master and slave were reading the same Bible to come to radically different political visions. The same thing happened in the nation as a whole. The sectional conflict is of course the most obvious example of the failure of the American synthesis to hold. Evangelical ebullience, forecasting a glorious future for a nation grounded in Protestant home life, perished along with 800,000 young men in what remains the bloodiest war ever fought by American forces. From the ashes emerged a new United States, a new family, and a new paradigm of education that would gradually eclipse the domestic ideals of the antebellum period.

Chapter Three

The Eclipse of the Fireside, 1865–1930

On June 14, 1863, Cornelia Peake McDonald sat with a friend on her porch in Winchester, VA, watching a battle progressing in the valley below. "We were yet on the outskirts, and could see the troops deploying, skirmish lines thrown forward and mounted men galloping from one point to another, batteries wheeling into position, and every now and then the thunder of cannon and the shriek of shell." The firing and shelling gradually came closer as the morning matured. "So they go, whizzing, screaming, and coming down with a dreadful thud or crash and then burst. We hold our breath and cover our eyes till they pass. I gather all the children in till the firing ceases." By noon the front was quiet, but only because a division of Confederates was sneaking around behind the Union forces to surround them. By three o'clock McDonald's Confederates had won the day, and Union troops were scattering. Some had wandered into her yard, where her two little boys, Donald and Roy, "seemed to forget the shells and were...running and catching the men as they passed and saying, 'I take you prisoner.'" Soon the Union forces were in full retreat, with the Confederates following after and raining shells down on them. "One battery from the hill opposite our house rushed down and through our yard, their horses wounded and bleeding, the men wounded also, and pale with fright." They were headed to a nearby Union fort, but it had already been taken by the Confederates. Bewildered, the Union men looked around for a safe perch, "and finally avail themselves of the only spot the shells did not reach, the angle of our house. I had retreated there with my children when the shots and shells began to fly so fast, and burst all around the house, and then as I sat on the porch bench, men came crowding in...They talk openly of being surrounded. The soldiers say they will stay and be captured."

Though these men were McDonald's mortal foes, she took pity on the wounded, one of whom had a shot ball lodged in his throat so that "the hard breathing as he struggled to keep the blood from choking him was dreadful to hear." Eventually crowds of Union soldiers had taken refuge around the house. Horse-drawn ambulances unloaded scores of wounded men. Horses "frantic with pain...were streaming with blood." And the Confederate barrage continued. Through it all "the children were leaning on my lap; I was holding my poor little Hunter. Roy and Nelly were perfectly composed, looking up at the shells as they flew over and came crashing down. Donald, poor little four-year-old baby, hid his face on my knee and sobbed." As evening approached, the soldiers began to wander into the home itself, until "before I knew it there were at least fifty men in the house." They spent the night sprawled on her floor and were greeted the following morning by "a column of grey coats!...They came up and halted before the door. I told an officer the Yankees were in the house; he asked me to send them out. I told them to go, and each one laid his musket down and marched sadly out."[1]

But though the South had won the day, it was not long before Union reinforcements forced McDonald to flee Winchester with her nine children for Lexington, where she stayed until the war's end. When word came of Lee's surrender "I felt as if the end of all things had come...The distress of the children was as great as mine; their poor little faces showed all the grief and shame that was in their hearts....I remember once glancing out at the window and seeing Donald who was too proud to show his concern to the family, walking up and down under the window with his fat little face streaming with tears, and wringing his hands in utter despair." The ensuing months were a terrible trial for the McDonald family. In her darkest moments, her husband dead, the pantry empty, she and the children starving, and the rent due, McDonald "felt that God had forsaken us....I had lost the feeling that God cared for us, that He even knew of our want." Slowly and gradually, thanks largely to the ministrations of a wide circle of friends and extended family, McDonald was able "to take care of my family till they were fitted to be of use themselves."[2]

From the Civil War to the New Deal

McDonald's is a tragic story, but many Southern women had it far worse than she. Tens of thousands of poor white widowed women and orphaned children, homes destroyed, roamed the Southern countryside looking for their next meal. Relief to the poor at the beginning of the war had been left to private charity, but as conditions grew worse and worse, state governments were compelled to intervene, often by providing pittance wages for women to make clothes or munitions for the troops. Despite such efforts, bands of women and children rioted almost daily throughout the South, many of them deeply resentful that the planter aristocracy was still eating well while they starved and their sons, husbands, and fathers died. Adding to the chaos was the steady self-emancipation of slaves, many of whom suffered the worst of privations during the war and were using their newfound freedom to take to the road. Formal home education under such circumstances was unthinkable, of course, but intense trauma is its own sort of education. Many Southern children of the war grew up permanently marked, be they blacks like Amy Penny, who suffered for years under brutal poverty and racism and concluded, "I think slavery wus not such a bad thing 'pared wid de hard times now," or whites like Ben Tillman, whose white-supremacist rhetoric captured for many Southern whites the bitterness their childhood experiences had bequeathed to them. In the North, while material privation was not nearly so severe, thousands of mothers, wives, and children mourned the loss of their men and were reduced to tears by the resonant strains of George Frederick Root's popular song:

> We shall meet but we shall miss him.
> There will be one vacant chair.
> We shall linger to caress him
> While we breathe our ev'ning prayer.
> When one year ago we gathered,
> Joy was in his mild blue eye.
> Now the golden cord is severed,
> And our hopes in ruin lie.[3]

Much more than the hopes of individual families lay in ruin after the war. For one, the war signaled the beginning of the end of Protestantism as the guiding national ideology. Cornelia McDonald was able to resolve her personal crisis of faith by remembering past trials that her Heavenly Father had seen her through. "With that remembrance came the resolve, 'Though He slay me, yet will I trust in Him.'" Millions of Americans on both side of the conflict turned to God during and after the War. But though personal piety was at a fever pitch, the result of the failure of the American synthesis to hold and the ensuing destruction of hundreds of thousands of lives and millions of dollars of property was a much more secular society. Both sides had defended their positions on slavery by quoting the Bible. Both had trumpeted the Divine blessing of their cause and were certain of God's providential hand in the war's progress. But in the end it was secular military strategy and munitions, not providence and Bible verses, which secured victory for the North. In historian Mark Noll's words, "The War, which had been fought on both sides to defend republican Christian virtue, led to a world in which that kind of virtue was not nearly as important as it had been before."[4]

Evidence of the secularization of postwar life is not hard to find. Children's books and advice books became much more worldly than their antebellum counterparts. In children's fiction, for example, static characters making clear choices between virtue and vice gave way to characters like Elsie Dinsmore and Harry Walton, who must navigate a complicated and dangerous world amidst bumbling and unreliable adults. The venerable *McGuffey's Readers*, far and away the most popular school textbooks throughout the nineteenth century, markedly shifted their emphasis after 1865 away from explicitly evangelical religion toward a vague, nondenominational Christianity, even erasing all anti-Catholic references. In the 1850s, children had practiced their penmanship on phrases like, "Fear God and keep all his Commandments" and "No man may put off the knowledge of God." By 1870 they were transcribing such gems as "Fortune favors the brave" and "Command all excellence." Evangelical religion lost its hold on the educated elite as Darwinism challenged the underlying Christian account of human origins, and the historical criticism of the Bible undermined confidence in its authority and reliability. The result was a split between liberal Protestants who stayed within the cultural mainstream by sacrificing the doctrinal content of their religion and fundamentalists who latched on to a narrow set of orthodoxies and fought as hard as they could against encroaching modernism.[5]

Other splits plagued the country as well. Many Americans struggled to reconcile their inherited "island community" value systems with the newly emerging national, corporate, urban-industrial complex rapidly constructing itself by commercial speculation and connecting its centers of production by miles and miles of railroad. Gilded Age America was experiencing the fragmentation of belief, as Christian moral restraint seemed more and more at odds with the marketplace consumerism that was emerging as the dominant cultural force. Industrialism brought all sorts of social fissures into the foreground as well: class struggles completely unnoticed by antebellum commentators such as Alexis de Tocqueville became a daily reality due to the extension of the wage-labor system in the cities and the consolidation of transport ownership in a few hands; racial tensions mounted to new heights in the North as homogeneous communities were replaced by ethnically mixed locales with limited material resources and in the South as redemption politics and supremacy movements sought to keep the black populace from realizing the fruits of liberation. "America in the late nineteenth century," says historian Robert Wiebe, "was a society without a core. It lacked those national centers of authority and information which might have given order to such swift changes."[6]

Caught up in the midst of all of these changes and transformations was the American family and its domestic education. Perhaps the most striking change in the family after the War was the dramatic reduction in children born to married couples. In 1800 a married woman averaged seven children. By 1900 the figure was down to three and a half, and by 1929 it had dropped below three, though it was even lower among middle-class families. The birth rate actually *rose* for families at the bottom of the economic scale. Families whose children went to work rather than to school had an economic incentive for having large families, and such families were the most likely to escape poverty as kids helped parents pay the bills and eventually even buy a home. The average middle-class family between 1850 and 1880 had fewer than three children, and the figure kept declining steadily until after World War II. In addition, middle-class mothers were spacing their children closer together, thus ensuring several decades of active life after the children had grown, if they and their children were fortunate enough to survive that long. For though birthrate declined after the Civil War, child mortality remained high until twentieth-century developments in public health and inoculation. Children under five accounted for 40 percent of all yearly deaths throughout the nineteenth century. And their mothers continued to die in childbirth as well. Many a nineteenth-century youngster was reared by a stepmother, some of whom lived up to the fairy-tale stereotype, making domestic education awkward and painful. Frank Klingsberg, born in 1883 in Kansas, was one stepson who "learned to be happy away from home."[7]

Why the dramatic birthrate plunge? Some historians have emphasized fears for the woman's health and safety in an era when gynecological problems were little understood and childbirth was dangerous. Others have stressed economic factors such as declining availability of arable land and middle-class anxiety for the financial futures of their children. Some others have noted the correlation between female education and birth rate. The higher a woman's educational level, the later she married, and the fewer children she had. Whatever the reason, the process was certainly abetted by the Victorian convention of the "sexless woman," the notion that the female sexual instinct was very minimal. But though the official Victorian view was that marital intercourse ought be rare and procreative in nature, the abundant trade in contraception suggests strongly that many people led double lives. Contraceptive devices were becoming increasingly available toward the end of the nineteenth century, so much so that when the Comstock Act was passed in 1873 to suppress obscene literature in the mail, contraceptives and aphrodisiacs were prominent among the seized contraband. By 1890 the average age of marriage began to decline, even for highly educated women. Why? The reason was contraception. Contraception, more than anything else, facilitated the transformation of marriage from an economic arrangement, whose main aim was to produce and rear children, to a dyadic relationship stressing companionship and intimacy. Victorian prudery made perfect sense when sex meant children. But after contraception became generally available, it slowly became acceptable to admit that women are sexual beings too.[8]

The slow shift in the understanding of marriage and family life from being grounded in the natural order of things to being a human arrangement meant to maximize individual well-being has a long and fascinating history with roots going all the way back to the Reformation's de-sacramentalization of the marital bond. But it got a big push from the Industrial Revolution. It has long been understood that one of the most dramatic and profound shifts caused by the industrialization of production was the narrowing of functions in the home. As productive work increasingly became located outside the home and designated as a male activity, women's domestic role shrank. Homemaking in the nineteenth century was still a full-time job, and it often made financial sense as well. A full-time housewife could improve the family situation by growing vegetables and tending animals, preparing food, taking in boarders, perhaps offering childcare or drawing, dancing, or music lessons in the home. But every year new labor-saving devices and industrial techniques took away more and more of a housewife's duties from her. The emergence of professional ideals and standards cut women off from skills they might have learned at one time from their husbands and fathers. Increasingly, practices from medicine to metallurgy were only taught to those training for the profession. And as farms increasingly focused on single crops or livestock that could make money, the farm work women did became less valuable since it didn't bring in any cash. Wood stoves, sewing machines, iceboxes, and, in the early twentieth century, electricity, indoor plumbing, canned food, and a host of other technologies had the dual effect both of de-skilling domestic labor and increasing the amount of grunt work a housewife had to do: factory-made cloth meant more clothes and linens to wash and sew; store-bought flour and sugar meant more demand from family members for baked goods; more furniture and bigger houses meant more cleaning. And to top it all, servants were getting harder and harder to come by since they were finding better wages in factories. In short, as one commentator explained it, "The creative process has been taken from the homes and lodged in machines and factories." And as more and more products became available, many of them previously unknown to most Americans, the mother's role increasingly shifted from production to discriminating consumption. She became the shopper, especially as door-to-door services gave way to shopping districts that required the consumer to take over the tasks of transporting and distributing goods, a process that increased dramatically with the spread of the automobile in the twentieth century.[9]

Industrialism changed the physical attributes of the home as well. We tend to think of suburbanization as a phenomenon of the 1950s, but in fact it goes back to the period after the Civil War, when middle-class Americans escaped the poverty and immigrant cultures of the cities by settling on the outskirts, with horse-drawn streetcars for transport and new schools for their kids. Homes got bigger, more private, and more lavish in design and furnishing. Industrialization made all of this possible by dramatically decreasing the cost of building and furnishing a home: cheap, standard-sized lumber, metal nails, factory-made windows and doors, furniture sold by catalogue, all brought to the consumer by rail or steamboat. Industrialism brought luxury to the American middle class on a scale never before achieved in human history.[10]

For the late nineteenth century, the domestic focal point of this luxury and civilization was undoubtedly the parlor. The parlor was the room that showcased a family's sense of style and leisure activities. Here family members assembled, entertained guests, and passed the time with games, theatrical productions, media consumption (like the faddish photographic viewers known as stereoscopes), singing and listening to music, writing letters, and, most popular of all, reading. A typical parlor had a center table surrounded by chairs and perhaps a sofa. The table was often covered with a deep-toned cloth and perched on a richly colored rug. On the table was usually a large family Bible, assorted books and magazines, and perhaps some photographs. The parlor was stuffed with other icons of civilization: portraits, scientific specimen, maps, wall hangings, a bust of a famous figure, perhaps an organ. But the most important parlor object was the family book collection. Books, as one historian noted, "brought learning home." The most well-to-do might have an entire library as a separate room, but even more modest middle-class homes had a decent collection of books in the parlor.[11]

Reading was the country's preferred leisure-time activity. By 1860, ninety-three percent of men and 91 percent of native-born white Americans could read. Most middle-class families owned several books, thanks largely to industrial techniques of bookmaking that dramatically increased the supply and variety of books while decreasing their cost. Late nineteenth-century Americans read a lot of history, biography, travel, geography, natural science and so on, anything that improved the mind. But their passion was really for novels. Novels were controversial because they were seen by many as unserious and even subversive or irreligious. During the first year of the Comstock Act's passage, federal authorities seized 100,000 books, many of them novels, which were deemed indecent (along with 200,000 pictures and 5,000 packs of illustrated playing cards). But the publishing revolution could not be contained. Books were everywhere. One commentator claimed that in the 1880s, "Goethe, Dante, and Shakespeare are read in the backwoods of Arkansas and in the mining camps of Colorado, in the popular 16 and 20 cent editions, by people who could never have afforded the books" in hardcover. But it wasn't just the classics. The country was blanketed with dime novels, especially mystery series, of which 1,551 new titles were published in 1886 alone. And their numbers kept rising. Three times as many books were published in 1900 as had been in 1880. In addition to books, millions of Americans subscribed to newspapers. Even rural families with no local paper would eagerly anticipate a shipment of recycled papers from friends or relatives in cities. And then there were the monthly and weekly periodicals aimed at niche markets of every sort. The variety and popularity was mind-boggling. In 1885 there were 822 foreign language periodicals published in the country. By 1913, the high-water mark, there were 1,323. Ten-cent "lowbrow" magazines accounted for more than 80 percent of all periodicals purchased, especially the new pictorials and mass-marketed women's magazines like the *Ladies Home Journal (LHJ)*. By 1903 the *LHJ* had over a million subscribers. By 1920 fourteen more magazines, some religious, most aimed at women, had secured at least a million subscribers each.[12]

So the Victorian moralist seeking to stock her or his parlor library had some difficult choices to make. Historians have noted gradual shifts during this period in parlor activities that dovetail nicely with larger changes in American society. Postbellum parlor activities were focused largely on the education and improvement of those participating. A family's religious life was often centered here, as father (or, increasingly, mother) would read from the family Bible and lead in prayers in the evenings. Most reading was communal: it had to be in an age when the only light in the room came from a gas lantern hung from the ceiling above the table. Parlor games and theatricals were explicitly educative activities. But gradually and slowly, parlor activities began to shift toward entertainment for its own sake rather than simply for its moral or intellectual payoff. The religious dimension gradually diminished as well, illustrated by the replacement of the organ and its hymns with the piano and its repertoire by famous European composers. And with the introduction of Thomas Edison's electric light in 1879, reading and other activities after dark could be done in private. (In 1890 less than 5 percent of homes in Muncie, Indiana had electricity. By 1925 this increased to 99 percent). As entertainment and physical comfort gradually replaced formality and intellectual culture, the parlor eventually gave way to the living room with its roaring fireplace (and later its radio) as the place where families spent their leisure hours. Parlors remained in many homes, and in some cases still do, but often as cold, uninhabited spaces with attractive but seldom used furniture. And the family Bible, if it still existed at all, rested there, imposing and unopened.[13]

But why? Why did the buttoned-down Victorian family of the late nineteenth century go casual? Why did it produce the Jazz Age and the flapper? Why, in the words of literary critic Malcolm Cowley, did women in the 20s "smoke cigarettes on the streets of the Bronx, drink gin cocktails in Omaha and have perfectly swell parties in Seattle and Middletown?" There are many possible explanations. First, as we have noted, the reduction of the birthrate and the closer spacing of children left middle-class women confined to domestic roles with little to do for decades of their lives. In the 1870s thousands of such "praying women" took their moral concerns into the country's streets and saloons, preaching to wayward men (and sometimes destroying property) in the name of home and mother. In 1874 alone there were 3,000 such "rum sieges" put on by the Women's Christian Temperance Union (WCTU) and similar organizations. Frances Willard, the WCTU's most charismatic leader, wanted her organization "to make the whole world Homelike," but in hindsight it is clear that the female suffrage for which she fought and the prominent public lives she and others led paved the way for a feminism that Willard and other straight-laced moralists like her would never have countenanced.[14]

Another possible explanation is that the Victorian moral consensus' exclusivity proved its undoing. Most late nineteenth and early twentieth century respectable Americans were extremely fearful of the country's growing minority populations: Catholics, African Americans, immigrants from Eastern Europe, and other groups. As these groups grew in number and political power the Anglo-American nativist simply could not maintain control of the country. But while there is no denying the pervasive racism of Anglo-Americans during this period, this explanation does not satisfy me, for it is clear that many of these excluded groups embraced domestic ideals very similar to those of the Victorians. Many immigrant groups brought with them assumptions about women's roles at least as conservative as those of middle-class America, and the threats of modern society ironically helped *strengthen* such ties to community and family for many transplants. Many immigrants assimilated quickly with the cultural mainstream. Catholics, for example, were just as quick as Protestants to embrace the parlor organ and to fill their homes with the other trappings of middle-class life as soon as they could afford to do so. The rebellion against Victorianism came not from outside but from within Victorian Culture itself. "The revolution in morals," said Cowley, "began as a middle-class children's revolt."[15]

It did so because, as we have seen, late nineteenth-century Victorianism was a culture without a core. The Civil War had left the country without a public religion, and as the decades passed many Americans lost their private religion as well. A society's moral codes can live on for a time without their underlying philosophical justification, but they do so as vestiges. And gradually, as new generations that do not believe or even understand the old justifications gain ascendancy, the morality itself is cast aside. This is partly what happened in the United States in the 1920s. Americans coming of age in that generation had been raised by parents who adhered to the moral codes of the past though many of them no longer believed in their underlying rationale. And as commentator William Phelps noted in a 1924 piece, "Skepticism in religion is, in nine cases out of ten, followed by skepticism in morals," only sometimes it takes a generational change for the implications to be worked out. As Darwinism and commercial materialism ate away at traditional religion, it is not surprising that children brought up amidst advertising celebrating this-worldly pleasures (and education equipping them to afford them) would find the Victorian code of respectability outdated and constricting. And when you add to this void the trauma of World War I and the national joke that was Prohibition, the roaring 20s begin to make sense.[16]

But it is actually more complicated than that, for many of the parents of Jazz-Age youth and a large number of their children as well were not religious skeptics or self-conscious bohemians. The industrial revolution had created a society of material abundance. Victorian Americans, like their antebellum evangelical forebears, wanted both material plenty *and* traditional morality. Unwilling to sacrifice either, they engaged in evasive culture wars, thinking that if they could censor immoral messages they could keep America pure. Rather than taking on the mass production of print itself, for example, they raged against comic strips. Mary Pedrik complained in 1910 that newspaper comics are "a carpet of hideous caricatures, crude art, and poverty of invention, perverted humor, obvious vulgarity, and the crudest coloring...which makes for lawlessness, debauched fancy, irreverence." This sort of thing could salve the consciences of adults and allow them to embrace both their nostalgia for the good old days and their own decisions to embrace consumerism so long as it wasn't explicitly titillating or subversive. But it could not win the day, for the underlying cause of the revolution in morals that was taking place was not the particular message of this novel or that magazine—it was the entirety of industrial capitalism that was systematically undermining the premodern village morality so many Americans still wanted to live by.[17]

There were two sorts of troubled conservatives. One set, often forgotten by historians, was the vast number of Americans who still lived a largely premodern life. As late as 1870 seven out of ten Americans still lived in small rural villages or on farms. The year 1900 was the first year in American history when there were more Americans living in cities and towns than on farms, but as late as 1917 there were still 32.5 million farmers, a figure that remained fairly constant until the 1940s. While some of these farmers were self-consciously progressive in their methods and mechanization, many were only a few steps removed from ancient farming practices going back millennia, and even if their techniques were industrialized, most of them still thought and believed like traditional farmers. To these people, a majority of the population until 1900 and a very large minority for long afterward, the newer permissive morality could be bewildering and frightening.[18]

A second group of conservatives had recently left the farm or the small town and were now entering the new industrial world of finance, business, education, and the professions. Many of these people carried with them nostalgia for the world they left behind, even as their own lives betrayed such loyalties. Harriet Beecher Stowe's lament, "My mother was less than her mother, and I am less than my mother" is typical of the sentiment that pervaded this dislocated group. They believed, in the words of a *North American Review* article from 1888, that if "the human race be cut off from personal contact with the soil" and "the healthful simplicity of nature" then "decay is certain." But they moved into the cities just the same. Some of them tried to keep faith with rural life by maintaining gardens or livestock (many suburban homesteads had domestic animals like ducks, chickens, hogs, horses, and cows roaming around the premises) or by taking up woodworking or other crafts. But these things were done more for their therapeutic value than for their actual productivity. As Jackson Lears has argued, such antimodern gestures "helped ease accommodation to new and secular cultural modes" by transforming the meaning of rural tasks. Craftsmanship, for example, "became less a path to satisfying communal work than a therapy for tired businessmen."[19]

Neither true agrarians nor citified Americans nostalgic for a lost agrarian past challenged industrialism itself. Rather, they sought to bolster the social order through symbolic acts. Was the family threatened? Petition the congress and pass a law creating Mother's Day! (first celebrated in 1914). Were the kids up to no good? Pass laws outlawing Jazz dancing! In the early twentieth century, government was on the side of the moral conservatives. This explains why American fundamentalism, a religious movement among Protestants explicitly directed against modernizing and liberalizing trends in society, did not mount a war against political institutions and figures. Fundamentalists of the early twentieth century did not leave the public schools. They did not homeschool. Why? Because the schools, like other government institutions, enforced Victorian morality and traditional religion. When schools did not do so, as in the classroom of John Scopes in Dayton, TN, the fundamentalists challenged the system and won. The Scopes trial is still too often seen through the lens of H. L. Mencken and *Inherit the Wind*. Such a perspective obscures the fact that Scopes and Clarence Darrow *lost* the case, and that for decades science textbooks around the country avoided the topic of evolution like the plague.[20]

Government, like conservative individuals generally, wanted a society that was both modern in its industrial productivity and traditional in its family life. From President Theodore Roosevelt down, government officials were aghast at the declining birth rate and soaring divorce rate among native whites while immigrants flooded the country and bred promiscuously. Roosevelt famously called these trends "Race Suicide." For 25 years, the U.S. Department of Agriculture under Liberty Hyde Bailey and successors sought to prop up native stock by encouraging rural family life through such programs as Homemaker and 4-H Clubs for the retraining of farm wives and children and by subsidizing subsistence homestead settlements in the 1930s. It was joined by organizations like the American Eugenics Society, whose dual goals were to get immigrants to use birth control and native whites to have bigger families. "Better Babies" and "Fitter Families" contests appeared around the country at county fairs and other venues, giving out medals and prizes to large native families based on hereditary "fitness."[21]

In fact, one could interpret many progressive and New Deal initiatives as efforts to save the traditional family from perceived disintegration. In 1911, for example, Illinois enacted the first statewide law giving government aid to widowed or abandoned mothers. In eight years, thirty-nine states had similar laws. Such laws were lobbied for and crafted by progressive minded activists who worried that "the good old-fashioned home has absolutely broken down." Aid was carefully earmarked only to "suitable homes," not to those where "inefficiency and immorality" reigned. If mothers did not attend church, kept a dirty house, or used tobacco products, they risked having their funding cancelled by social workers who monitored them. American families came under increasing scrutiny by all sorts of government agencies rallying around the cry, "save the family!" Prohibition agents investigated at-home drinking. Feminists pushed for population control even as maternalists sought to use government to outlaw contraceptives. Marriage and divorce laws were stiffened. Child-labor and compulsory school attendance laws, passed by nativists in the Republican party and affirmed by their allies in state courts (and at first not supported by school people or Democrats), sought to "keep parents from exploiting their children economically" and to ensure their attendance at schools that would inculcate middle-class Protestant notions of family life. The U.S. Children's Bureau, formed in 1912, pushed for higher wages for working men so that wives could stay home and for providing scientific training for mothers so that babies would be born healthy and raised well. Vocational training in schools was given federal support through the 1917 Smith-Hughes Act in response to criticisms by reformers like Florence Kelly, who complained:

> The schools may truthfully be said actively to divert the little girls from home life... For the schools teach exactly those things which prepare

girls to become at the earliest moment *cash children* and *machine tenders*: punctuality, regularity, attention, obedience, and a little reading and writing—excellent things in themselves, but wretched preparation for...homemaking a decade later.[22]

New Deal work programs providing a "family wage" to men (and not to women), the Social Security Act that guaranteed pensions for jobs associated with the male breadwinner, home ownership programs that revolutionized housing finance through long-term loans (with low down payments and interest rates) and spurred new housing construction through federal insurance for developers—all of these were government initiatives to save the two-parent, working father and stay-at-home mother, family. Reformers might differ on specific remedies, but they were all united by the belief that to save America from dysfunctional homes "it was necessary to expand the state's supervisory and administrative authority." All this energy produced what historian Morton Keller has called a "revolution in public philosophy" about the relationship between parents and kids. By the early twentieth century, government was taking a much more active role in overseeing and regulating parenthood, and it was doing so, it must be stressed, to *save* the traditional family. Faced with large-scale breakdown of the stable two-parent family, Americans turned to their government to solve the problem. Family courts were created to deal with parental neglect, adoption, juvenile delinquency, and custody after divorce. "Manual training" programs in public schools, houses of refuge, reform schools, YMCAs in cities for disoriented rural migrants, penitentiaries, and all sorts of other institutions were modeled on the family and implemented as surrogates for those whose own families were dislocated by industrial change. Whereas past generations of Americans had looked to the family to keep the nation strong, it was now up to the nation to save the family through the interventions of professional expertise.[23]

Parents and Professionals at Home and School

Home-saving reformers were especially enervated by the prospect of professionalizing parenthood itself. During the 1880s and 1890s "Domestic Rationalizers" dressed up old ideas in scientific lingo to create the science of "homemaking," a word coined at this time. The homemaker would apply rigorous scientific techniques to her sphere just as her husband was doing in his factory, laboratory, or office. One of the main goals of the many new home magazines enjoying such dramatic popular success was to help educate women in proper homemaking tastes and techniques. Turn-of-the-century progressives hoped that by professionalizing motherhood they could make the domestic vocation attractive to a new generation of educated women. Their efforts extended into many domains previously governed by inherited tradition and folk wisdom. Childbirth began to be viewed not as a natural phenomenon presided over by women in the home but as a medical situation to be treated by male doctors in a sterile hospital setting complete with modern pain relief and scientific procedure. Housework was rendered scientific through labor-saving devices, motion studies, and increased reliance on industrially processed foods. Mountains of literature advised parents on the proper techniques for the feeding of infants, toilet training, personality development, and every other conceivable topic pertaining to child-rearing. Many of the women churning out this expert advice were not mothers, but they were college-educated. And by the 1920s a fairly large number of American women were attending college. These women, as historian Barbara Beatty explains, "in lieu of listening to their mothers or other traditional sources of information about child rearing, turned to women like themselves." They wanted a textbook. And they got it.[24]

The book was called "home economics." More than just a textbook or a course of study, it was a movement. Its founders were convinced that industrialism had changed the world and that if the family were to survive, it would have to adapt to the new conditions. Thus, they tirelessly preached a new sort of domestic education, an education conducted not by parents for children in the home but by experts on parenting itself. A host of institutions joined in the effort: kindergarten and infant schools; public schools offering new programs in parent education, home visits, and domestic science; universities sponsoring child study projects, home economics courses, and "practice houses;" and of course American businesses, whose advertising relentlessly targeted the American mother with pitches explaining why their products were safer, better, more modern than the homespun and scratch-made fare of yore. Home economist Dorothy Baruch, Assistant Professor of Education at Whittier College, summarized the orientation animating all such efforts:

> Parenthood should be a trained profession, not a hit-or-miss affair left to instinct alone.... Fortunately we as parents can become educated in

our work of parenthood. All over the United States organizations of many types are making available courses and lectures and guidance in reading that will help men and women toward finer parenthood.[25]

But what of parents who didn't read the textbook? Underneath the pro-family rhetoric of the home economics, kindergarten, and other maternalist movements runs a deep current of distrust of actual parents, especially poor immigrant parents. Many activists were so wedded to their theories of child development and psychological adjustment that they considered parenting not done according to scientific principles tantamount to child abuse. Few were so outspoken about their scorn for uninitiated parents as progressive educator Caroline Pratt, but she aired what many were thinking when she declared that there are "no bad children, only bad parents." She wondered why so many parents "gave so little of themselves to their children," but thought it was perhaps just as well, for many children would be "better off without them altogether." For Pratt and other home savers, an institutional mediation like the nursery school was a child's "first step" toward "emancipation from the home." It was this distrust of poor mothers that had led home education advocates like Elizabeth Peabody into the Kindergarten movement in the first place (In the 1830s Peabody had been writing about *Family School*. By 1870 she was publishing *The Kindergarten Messenger*). And by the 1920s, Pratt's skepticism about the private family's suitability for raising healthy children was shared by the great majority of women leading child study and early childhood education organizations.[26]

* * *

One of the key features of the professionalization of any field is specialization. The new doctrines of parenthood held that parents should specialize in the emotional and psychological development of their children while intellectual training should be left to other professionals. After the Civil War and for decades thereafter the number of children attending public school and the duration of their tenures rose dramatically. Public high school enrollments especially surged, almost doubling every decade from 1890 to 1930. By 1935, of all American youth, 40 percent were graduating from high school. The dramatic increase in students at all levels created a true administrative crisis that was met by the new scientific spirit permeating the educational profession. School leaders studied, experimented, and collaborated their way

toward crafting a school system that could handle all of these enrollees and do so with efficiency and beneficent social outcomes. Even as the family was becoming more intimate and informal, the school was growing larger, more impersonal, and further removed than it had been from home life, taking on more and more of the functions parents had historically performed. New programs in health and hygiene, vocational training and guidance counseling, physical education, and especially extracurricular offerings like school sports and socials, were turning the school into an incubator for peer culture and adolescent identity-formation.[27]

But why did parents so willingly relinquish their authority in these matters? Why did almost all of them send their children to public schools, parochial schools, or private academies? In light of the principled rejection of institutional schooling by many homeschoolers today, this question must be answered. First of all, middle-class Americans wanted their kids in school because the schools fitted perfectly with their notions of propriety and their aspirations that their children stay middle class. Large numbers of immigrants and blue-collar workers, many aspiring to achieve the middle-class dream themselves, felt the same way. Labor unions, for example, supported public education because it both kept children out of the work force (thus keeping wages high and jobs secure for men) and provided a means of upward mobility for working-class children (though how much mobility was actually permitted or achieved is the subject of fierce historical debate). And with laws prohibiting child labor and requiring school attendance, what else was there for kids to do anyway?[28]

Some immigrant parents, such as southern Italians who settled in New Haven, CT, themselves largely illiterate, had little use for schools and only wanted their children to work to contribute to the family wage. In cases like these, truant officers were hired to enforce compulsory attendance laws despite the parents' wishes. Italian children so forced into school quickly developed a taste for American fashion and popular culture and began to understand the value of school as a means of escaping the ghetto. Historian Stephen Lassonde has shown how the second generation learned to forget Old World customs like arranged marriage, deference to elders, and manual labor. School, and the jobs it opened up, taught immigrant children to be autonomous. Many immigrants could not bear to have their children completely Americanized, and Catholic education became the preferred mechanism for adapting to the new American situation without sacrificing their most cherished traditions. Sociologist David Baker has shown how urban America experienced a dramatic rise in Catholic school enrollment between 1880 and 1930. In some cities nearly half of the school-age children were in Catholic schools. Baker notes that Catholic officials could have chosen other educative models, but they "conspicuously borrowed an institutional model of mass schooling from the public sector to school some of the least-educated populations in American society, such as immigrants from Italy and Eastern Europe."[29]

But what of the American heartland? Why would the rural farmer or small-town resident willingly send his or her child off to school? Historians Claudia Goldin and Lawrence Katz have uncovered several overlapping reasons. They note that the growth in schooling, especially secondary schooling, was more dramatic in the American interior than anywhere else in the nation, and it did not occur by external imposition of state or national government. It was a grassroots movement. According to Katz and Goldin, Americans in what we now often call the "red" states embraced high schools for six reasons. First, they recognized that high school was a smart investment. In the early twentieth century, every year a child spent in high school raised that individual's potential earnings by about 12 percent. High school graduates earned on average twice as much as those who did not attend school. Secondly, rural areas recognized that if they did not provide secondary education for their children, the children would leave for the cities. They built schools close by to keep their kids local. Ironically, however, this very schooling proved excellent training "for those who wanted to leave the region and give up farming." Thirdly, small-town America embraced secondary schooling as a sort of "intergenerational loan," whereby older citizens would pay for the education of the next generation in the expectation that conscientious youngsters would grow up to take care of their elders. Fourthly, a relative cultural homogeneity and classlessness meant that there were not many elements within the population of small-town America that would protest against an institution that brought everyone together and solidified community loyalty. Fifthly, there was not much else rural children could do, given the mechanization of farming and the relative scarcity of other employment opportunities. Schooling often became the only hope a farmer had of securing a viable future for his offspring given the declining prospects of independent farms across the nation. And finally, the high school benefited from and contributed to the rich bonds of community and mutual reciprocity that gave many small towns their distinctive character. High Schools (especially their sports teams) became embodiments of local pride. Given such powerful reasons, it is not surprising that the public school became so popular for bread-and-butter Americans. It also helps explain why schools became such a prominent stage of conflict when they began to be used to challenge some of the cultural values of these regions. But in the early twentieth century, schools were not in the business of challenging majoritarian cultural norms. They were there to reinforce them. Religious people especially, Goldin and Katz found, were far more likely than those without church affiliation to send their children to high schools.[30]

That is not to say that there were never conflicts between the home and the school. Trouble erupted sporadically throughout the nineteenth century, especially in urban centers where large populations of immigrants objected to English-only instruction, as Germans did in San Francisco and Chicago. Catholics objected to Protestant Bible readings and anti-Catholic textbooks in Philadelphia, New York, Cincinnati, and elsewhere. In some cases these conflicts led to rioting, destruction of property, even homicide. Disagreements of a more interpersonal nature, such as those over truancy complaints, corporal punishment, or homework, were usually solved informally, person-to-person. But as school systems expanded and formalized, face-to-face encounters between parent and teacher were gradually replaced by report cards, scheduled conferences *at school*, and the school-led management of parental feedback in the form of Parent-Teacher Organizations.[31]

School leaders quickly discovered that it was more efficient and effective to manage parents in groups than individually. As Mary Harman Weeks, vice president of the National Congress of Mothers (later named the National Congress of Parents and Teachers), the organization that successfully coordinated the nationwide system of local and regional parent-teacher associations, explained, "Principals find the parent-teacher circle an excellent means of reaching all parents effectively when some general condition needs changing, when public sentiment in the district needs rousing, or when they wish to make certain courses effective which do not seem to take hold." The one-sidedness of communication here is obvious, and it stems from a straightforward source. In the twentieth century, teachers and administrators had finally secured for themselves at least a bit of the respectability associated with professionalism. The argument was constantly made that the teacher was like a surgeon, that laypeople could not possibly understand the full complexity of expertise involved in the esoteric task of teaching. Parents must, in

Arthur Perry's words, simply "have faith in her training and professionalism." In the early twentieth century, with the authority of science riding high and most parents inclined to deference, educators largely succeeded in corralling dissent. As their position solidified, teachers and other school people became less defensive and assertive of their authority and opted for a more cooperative spirit. "It was never intended," asserted educator M. A. Cassidy, "that the school should supplant the home in child training... nor is it desirable that the teacher should supplant the parents.... They should be co-workers." And so long as the school maintained good discipline and reinforced public piety, most parents cooperated willingly. Indeed, parental distrust of progressive novelty and preference for traditional curriculum and pedagogy is an important and often overlooked factor in the failure of many progressive ideas to take root in American schools.[32]

Most parents, then, were comfortable with the new division of labor. Indeed, it was the morally conventional middle-class family with the stay-at-home mother who was typically the most ardent supporter of public schools. There are exceptions. Turn-of-the century conflicts between a parent who wanted to teach a child at home and the state's compulsory attendance laws led to a smattering of court cases. In Massachusetts (*Commonwealth v. Roberts*, 1893) the court found that state law permitted instruction "by the parents themselves, provided it is given in good faith and sufficient in extent." In Indiana (*State v. Peterman*, 1904) the court found that a private tutor was legitimate since the state attendance law did not extend to the "means or manner" of the education provided. In Washington, in contrast (*State v. Counort*, 1912), the court found that home instruction did not count as a "private school" and was thus illegal. Similarly, in New Hampshire (*State v. Hoyt*, 1929) home tutoring was rejected both because of inadequate socialization and because regulating it would unreasonably burden the state. And in *Parr v. State* (1927) the Ohio Supreme Court upheld the conviction of a homeschooling family, stating that "the natural rights of a parent to the custody and control of his infant child are subordinate to the power of the state." But in Oklahoma (*Wright v. State*, 1922) home instruction was allowed so long as it was done in "good faith" and provided training "equivalent" to that of a formal school. The Supreme Court's landmark decision in *Pierce v. Society of Sisters* (1925), while famously holding that "the child is not the mere creature of the state" and forbidding the state from prohibiting private schooling, nevertheless gave no clear sanction to domestic education. The court's opinion explicitly stated that

> no question is raised concerning the power of the State reasonably to regulate all schools, to inspect, supervise and examine them, their teachers and pupils; to require that all children of proper age attend some school, that teachers shall be of good moral character and patriotic disposition, that certain studies plainly essential to good citizenship must be taught, and that nothing be taught which is manifestly inimical to the public welfare.

The question of whether a home school is a school was not answered by the Supreme Court then, nor has it been since.[33]

Prior to *Pierce* there had also been a few cases in various states involving challenges to compulsory school legislation by parents who wanted their children to work. In such instances courts almost always ruled against the parent. In *State v. Bailey* (1901), for example, the Indiana court upheld compulsory education because "no parent [has] the right to deprive children of the advantages" provided by an "enlightened and comprehensive system of education." But such challenges were rare. Whether challenging the very legality of compulsory schooling or simply trying to carve out space within it for home instruction, legal challenges were so infrequent that they only reinforce the general point that the overwhelming majority of Americans willingly and eagerly embraced formal schooling.[34]

The Home and Informal Learning

The home had thus ceded to the school most of the responsibility for formal intellectual instruction. But parents had a lot of educating still to do. Primarily, especially in the minds of reformers and the middle class, the home was the place where proper manners and morals ought to be imparted. In the words of one parenting manual:

> There must be constant home-training in the art of good behavior; and this children have a right to expect and demand of their parents. They must be taught how to enter and leave a room; how to bow, walk, turn, sit, rise; how to introduce people to each other; how to behave at the table, and, in a word, how to conduct themselves under the varied circumstances of life.

Another manual advised parents, "at the table a child should be taught to sit up and behave in a becoming manner, not to tease when denied, nor to leave his chair without asking." Though the emphasis here on manners may seem odd today given the dominant cultural preference for the casual and insouciant, these writers knew that (in John Dewey's words) "manners are but minor morals." It was the parents' job, especially the mother's, to create through diligent surveillance and example what David Reisman famously called "inner directed" people "whose conformity is insured by their tendency to acquire early in life an internalized set of goals." This internal moral compass would impart stable character traits no matter what new and challenging circumstances a dynamic society might present. The key to social and economic progress, thought many Americans of this period, was "the disciplined, autonomous self, created in the bosom of the bourgeois family." And everything about the family—dress, decorum, etiquette, even housing design, décor, and landscape—conspired to create that self.[35]

In addition to this moral task, many late nineteenth- and early twentieth-century families brought other sorts of learning into the home as well. Skills that were not being taught to children at school were often imparted at home, especially artistic endeavors like drawing and music lessons. Many children with special needs were cared for in the home, often by "visiting teachers" making home visits to offer remedial instruction to what were then called "defectives," orientation for recent immigrant families, and motivation for delinquents. Part social worker, part teacher, part parent, the visiting teacher's job was to "interpret the school to the home" and the "home to the school" by being a liaison for both.[36]

By far the most common form of education in the home during this period, however, was reading. We have already noted the dramatic rise of literature production and consumption during this period. While traditional practices of intensive and repetitive reading of authoritative texts (especially the Bible) continued, the flood of new material meant that much that people read was approached in a more cavalier manner. Group-reading continued to be popular, as the following anecdote of Rose Cohen, a Jewish girl growing up in New York in the 1890s, illustrates. Cohen often rented books for a few pennies from a local soda-water dealer. She would take the books home and read them aloud to her immigrant mother and siblings. "What a happy two weeks we spent!" Rose recalled upon renting Dickens' *David Copperfield*, "With what joy I looked forward to the evening when after supper we would all gather around the lamp on the table and sister or I would read aloud while mother sewed and the little ones sat with their chins very near the table...For just to read became a necessity and a joy. There were so few joys." At first Rose's mother had disapproved of the reading, but she soon became just as engrossed in the stories as her children. But Rose's father was never pleased:

> Father did not take kindly to my reading. How could he! He saw that I took less and less interest in the home, that I was more dreamy, that I kept more to myself. Evidently, reading and running about and listening to "speeches," as he called it, was not doing me any good. But what father feared most was that now I was mingling so much with Gentiles and reading Gentile books, I would wander away from the Jewish faith.[37]

Rose Cohen's story introduces a new theme in home education. Books, which were consumed by the millions at home, could challenge as well as reinforce the other messages being taught there. Many Victorians of course had their views cemented by a steady stream of literature that affirmed ideals they already had. But some discovered alternatives. Ella Reeve Blor, daughter of a conventional New Jersey druggist, was introduced by her eccentric great uncle to Darwin and the agnostic apologist Robert Ingersoll. She later became a socialist and union advocate. Many immigrant children internalized the norms and language of Anglo-American culture through their reading, and many others acquired the tools to critique it the same way. Reading gave a feminist such as M. Carey Thomas access to "male" knowledge and male role models that strongly influenced her adult career in higher education and politics. Then, as now, girls read "boys' books" at least as much as boys did. Female autobiographies of the time are much more likely to talk about books and reading, and surveys of reading habits from the early twentieth century to the present have consistently found women to be the predominant readers of fiction. Much of this reading might be denigrated as escapist, but therein lay its significance. A young Victorian girl who grew up escaping to foreign lands, identifying with an adventurous male hero, or sympathizing with the plight of the urban poor might eventually become a very different person than her parents would have wanted. The right book at the right time can change a life.[38]

Home Education on the Margins and for the Marginalized

Despite the phenomenal growth in school attendance during this period, there were always some families that continued to provide formal schooling for their children at home. For some this was a practical necessity given the frail health of a child or geographic isolation. In rare cases ideology contributed to the decision. Many Mormons in Utah, for example, taught their children in homes during the early years of settlement when polygamy was still widely practiced. Brigham Young himself had forty-six children. In his "Lion House," twelve wives lived together with nineteen daughters and eight sons. School was kept in the basement of the home by one of the wives, Harriet Campbell Cook, and private tutors were brought in to offer enrichments such as music lessons.[39]

A few wealthy families continued to use in-home tutors. The childhood of journalist William F. Buckley, Jr. serves as a fine example of what home education could become if one had enough money. Buckley's father was a second-generation Irish immigrant who had made a fortune from Mexican oil. He created for his ten children an ambitious system of home education involving a host of tutors in music, Latin, mathematics, rhetoric, and foreign languages (Buckley spoke fluent Spanish and French before he learned English). Due to the oil business, the family often lived overseas. From 1929 to 1933 the children were educated in their four-story house in Paris. There was a French tutor on the fourth floor, a Latin teacher on the third, an English teacher on the second, and a music teacher on the first. Every hour the children would pass one another on the stairs leading to the various floors for their next lesson. When the war forced the family to return to their estate in Connecticut, the children were immersed in a world of home instruction by a host of adults. In his autobiography Bill Buckley fondly remembered:

> We were superintended by Mademoiselle Jeanne Bouchex and by three Mexican nurses; fed and looked after by a cook, a butler, and two maids; trained and entertained in equestrian sport by a groom and an assistant, making use of Father's eight horses; instructed in piano by a twenty-three-year-old New Yorker who came and stayed with us three days of every week, giving us each a lesson every day on one of the five pianos in the house; and in the guitar or banjo or mandolin (we were allowed our pick) by a Spanish-born violinist who traveled once a week from Poughkeepsie.[40]

Of course, most home-taught children did not have the advantages of the Buckleys. Most learned this way not by choice but simply for lack of other options. In the western territories and states, schools had a very difficult time keeping up with population shifts. In Nebraska, for example, population swelled due to abnormally high rainfall in the 1870s. In one year, though 12,000 students were added to the rolls, only six new teachers were hired and nineteen new buildings constructed. In situations like this, parents picked up the slack. "On balance," notes historian Elliott West, "frontier home education was better than adequate." South Dakota homesteader Theodore Jorgensen's autobiography provides a vivid example of the sort of schooling that took place throughout the upper Midwest. "Since there were no opportunities" in the early 1920s "for me to go away to high school," he notes:

> our parents decided to have school in our own home. They invited a neighbor girl to come too, and she commuted three miles by horseback.... They invited other youths of the community to come to our home school. One girl rode seven miles from her home and back each day through the cold winter months. Four boys,... who lived too far away to commute, hitched some horses to a shack on skids, moved it to our farm, lived in it for the school year, and then moved it away. One girl, whose home was at some distance, lived with us and earned her way helping with the house work. All of the school activities took place in our home which had two small rooms used for classes, a loft where the three boys slept, and a lean-to which served as a kitchen and dining area and when one end was curtained off also served as the girls' bedroom. Our parents slept on a bed in one of the class rooms. Wonders can be accomplished when the need arises.

Jorgensen recalled his mother boasting "that no other high school in the whole United States could possibly have had students going on to earn more Phi Bet Kappa keys per capita." Jorgensen himself went on to earn a Ph.D. in physics at Harvard and to work on the Manhattan Project.[41]

The typical frontier house school was usually quite modest in curriculum and apparatus. The family Bible was often the first text from which alphabet and reading were taught, and the dirt floor might become the chalkboard on which letters and numbers were scratched. "Mother guided and directed somewhat the children as they learned to read" recalled one pioneer child, "Her mind was filled with the lore of the Bible as she had learned it when a child, with Mother Goose rimes, with words and music of gospel hymns and popular songs, with the sayings of Poor Richard." Poor families squeezed lessons into whatever time was available: Lucinda Dalton's father taught her in the evenings when he returned from digging in California; Bennett Seymour's father taught him and nine neighborhood children around the table by candlelight. Wealthier parents like railroad official's wife

Elizabeth Fisk could devote more attention to the task. After morning housekeeping chores were finished, the children would have an hour's study followed by recitations, writing exercises, and sewing practice. Then the kids would recess while she prepared supper. She and many other western settlers relied heavily on the mail to provide the latest publications from the east. Library associations and reading rooms sprang up quickly in frontier towns so citizens could share the literary bounties they brought with them or mail ordered. Thanks to such efforts, literacy rates on the frontier remained high, by some estimates even higher than New England's. Visitors from the east frequently reported observations like this, "It was a perpetual surprise to me to hear girls whose whole life had been spent on the plains or in the backwoods talk of Longfellow and Bryant, Dickens and Thackeray, Scott and Cooper."[42]

Of course, frontier children, like rural children throughout the country, learned about more than books at home. An Iowa farm boy recalled, "It required stern military command to get us out of bed before daylight, to draw on icy socks and frosty boots and go to the milking of cows and the currying of horses." Another remembered that, "I was the coal breaker for our family. Each day, after the chamber work for the cows and horses was done, I had to break the coal to be used for the next twenty-four hours. I was taught how to do everything around the house as well as to mend my own clothing." While children's work was contracting in the rest of the United States, on the frontier it was expanding. Children cleared land, worked the garden, hunted, and trapped. Girls as well as boys did this work, though traditional gender roles tended to surface as the homestead became more stable. For, despite the rugged living, most homesteaders came west not to escape civilization but to reproduce it. That's why mothers made their homes into schools "for lessons in refined living," bringing with them on the trail books, works of art, a clock, dishes, even parlor organs. Many a poor farmhouse featured a parlor organ in the main room, symbol of a civilization parents were trying to pass on to their children. The rituals of music playing and singing bonded families to the culture they had left back east but had not abandoned.[43]

This longing for civilization helps explain why frontier families established public schools as fast as they could. By the 1880s, public education was more accessible in the West on average than in the rest of the country (though attendance was spottier, especially during peak times in the agricultural calendar). It also helps explain the earliest example of formal home schooling curriculum. As we have seen, families that didn't have access to schooling were nevertheless eager to pass on the knowledge and manners of modern society to their children. Though too isolated for brick and mortar schools to reach them, a school-in-a-box certainly could. In 1905 Virgil Hillyer, headmaster of the Calvert School in Baltimore, MD, hit on the idea of offering his school's curriculum to local parents who, for whatever reason, could not enroll their children at Calvert. His teachers transcribed their daily lessons, and the material was mailed to subscribers. Within three years, word-of-mouth and well-placed advertising (especially in the *National Geographic*) gained Calvert ninety-five students in thirty-five states and eight foreign countries. By 1910 the enrollment was 300. By the 1930s the Calvert curriculum was being shipped around the world to students in more than fifty countries, many of them children of American missionaries. In the 1940s it was adopted by the U.S. Department of Defense for the children of military personnel in Japan and Korea. A 1944 story in *Time* magazine doubled the enrollment, which has continued to grow to the present day. About 11,000 homeschoolers were using the curriculum in 2006.[44]

Calvert was not alone in the business of providing correspondence education. A few religious groups offered correspondence programs similar to Calvert's, the most enduring of these being the Fireside Correspondence School, founded by Seventh-Day Adventist educator Frederick Griggs in 1909. The name was later changed to the Home Study Institute (HSI). In 1947 its K-12 programs were approved by the state of Maryland, and by the 1970s HSI was enrolling over 3,000 students a year. In 2006 the name was changed again to Griggs University and International Academy. By that time, K-12 enrollment had dropped a bit to around 2500 students.[45]

While Calvert and Griggs are notable for their longevity and curriculum designed for children, at the time of their founding they were only two of hundreds of mail-order home-study programs made available by entrepreneurs of all sorts, from established colleges and universities to private companies and enterprising individuals. Some, like A. A. Berle's popular *Self Culture* volumes, were pitched to parents as a means of advancement for their children in the home. But the great majority targeted adults themselves, tapping into the deep American desire for social and economic advancement. A pioneer in this regard was the Society to Encourage Studies at Home, founded in 1873 by Anna Eliot Ticknor with the aim of offering women robust courses in modern languages and literature, science, art, and history in the convenience of their own homes. Some of the most prominent ladies of Boston charitably offered their services as "correspondents" to women around the country, who first were issued a pamphlet with very specific directions on proper note-taking, and then shepherded through a guided reading course consisting of the same literature men were studying at colleges. The courses were available to any woman over age seventeen who could pay the two dollar fee covering shipping and overhead. Between 1873 and its closing in 1897, the Society provided over 500 society women from Boston and New York the opportunity to correspond with 7,000 women around the country who could get a college-level education in no other way.[46]

By the time of its closure, the Society had seen its methods being imitated by dozens of similar groups, perhaps the most influential being the adult education home-study programs of the University of Chicago, the University of Wisconsin, and the Chautauqua University. Joining these nonprofit organizations were more aggressive profit-making companies like the International Correspondence Schools (ICS) of Scranton, PA, founded in 1891. Begun as an effort to advance the careers of miners, by 1930 it had mailed curricula for over 370 courses to the homes of over four million Americans. Most of its clients were working-class Americans with little or no formal schooling aspiring to escape the drudgery of manual labor and factory work by taking the ICS' very practical courses in such fields as engineering, bookkeeping, drafting, and other white-collar occupations. The education they received from ICS gave some who had lived on the margins of the American dream for generations just enough know-how to step over the threshold into the middle class.[47]

ICS's rapid rise of course spawned imitators. By 1926 there were over 350 correspondence schools in operation around the country offering courses in every conceivable subject or field to approximately two million Americans a year. But so many of these schools engaged in questionable business practices or outright scams that by the late 1920s the entire industry was facing a public relations crisis. To rectify the situation and salvage their brand identity, nine of the top schools offering home-study curricula formed an umbrella organization, the National Home Study Council (NHSC), to police the industry and establish quality standards. Some efforts were made to curb the sensationalistic print advertising, overly aggressive door-to-door salesmanship, and ruthless collection of funds from the 90 percent or so of clients who did not complete the programs, but the NHSC was never willing to discipline itself as it ought to have done. Federal and state governments stepped in, responding to increasing criticism of the industry, with regulations that put most of these companies out of business. By the end of the 1930s the popularity of correspondence education had declined significantly. Lawsuits and government regulation played a role in this, but far more consequential over time were higher attendance rates at formal schools and more rigorous credentialing requirements for such careers as engineering, law, accounting, nursing, and secretarial work—jobs that had been the home-study industry's bread and butter.[48]

* * *

A final theme in the continued use of the home to educate at the turn of the century is the treatment of marginalized people. Progressive reformers tried many innovative experiments using the home to deal with the serious issues of the urban slums. Some college-educated women (and men) took the middle-class home into the cities, founding what were known as "settlement houses" in urban centers across the country. Jane Addams' Hull House, founded in 1889 in Chicago, is the most well-known, but by 1910 there were over four hundred settlement houses in cities all across the country offering all sorts of educative programs and other social services to the urban populations they served. Though Hull House became famous (or infamous, depending on the commentator) for its progressive politics and programs, most of these settlements were concerned largely with Americanizing immigrants through religious conversion and inculcation of Victorian norms of domestic life for women and work habits for men. And when blacks began to replace white immigrants in many of the settlement neighborhoods, settlement leaders tended to leave. By the 1930s the programs of those few houses that remained had been largely taken over by government schools, social workers, and welfare agencies. But for a time the settlement houses were a unique example of how a home might be used to educate and assimilate families who didn't live there.[49]

Another reform effort tried to go about things in the opposite fashion. Rather than bring the home to the slums, children's aid societies across the country sought to bring the slums to the home by placing orphaned urban children in rural farming families. The idea was the brainchild of Charles Loring Brace, whose New York Children's Aid Society became famous for its "orphan trains" shipping children from the streets of New York City to the heartland. Over a 75-year period the NYCAS placed out 200,000 children. Brace's aim was to place children of failed parents with a "virtuous and Christian family" where they might imbibe the work ethic and sober living of the rural populace. Brace himself had little sympathy for what he called the "stupid foreign criminal class" who were the "scum and refuse of ill-formed civilizations," so he felt no compunction against breaking up intact families: many of his orphans were not orphans at all. When the families of origin tried to reclaim their children from the society, they were rebuffed. Catholic and Jewish religious leaders were especially disturbed, for it seemed to them that the society was mainly interested in rescuing children from Catholicism and Judaism. As for the children, the results were mixed. Certainly many children benefited from the new home environments in which they were placed and went on to lead prosperous lives. But large numbers of children were treated like chattel. Many of the host families only took children for their work value, and many harbored strong prejudices against Slavs, Jews, Poles, Italians, and other ethnic groups outside of the Anglo-Teutonic majority.

By the turn of the century, Brace's organization was in decline. Criticisms of his methodology had turned much of public opinion against him. Many states passed laws prohibiting the shipment of children across state lines, thus encouraging more local children's aid activity. Philadelphia, for example, had two organizations that placed out 5,400 children to rural Pennsylvania homes between 1880 and 1905. Such groups shared Brace's underlying conviction that "the best way to fit a child for an active, industrious, wage-earning life is to place it in an active, industrious, wage-earning family," but the smaller scale and local placements allowed for better screening of host families, thereby curbing some of the worst abuses. But gradually the orphan train experiment was abandoned. Compulsory education laws made farmers less interested in orphans as a source of cheap labor. A new professional class of child savers was increasingly turning against the indenture system, favoring an approach that would keep children with their biological parents in an effort to save not just the child but the entire family. In 1929 the New York society stopped running orphan trains altogether. But the preference for family placement over institutionalization lives on in the foster care system of today.[50]

And finally, there were always some homes that stayed beyond the reach of child-labor and compulsory education legislation. Many poor mothers continued to keep their daughters home to be "little mothers" to younger children and to help around the house. Working-class children sometimes reported to horrified reformers the pride they felt contributing to household survival through their labor. One child in a mill town recalled how she "just wanted to work and make some money... I didn't want to go to school." She and many of her peers were able to meet this goal through their mill work, but just as importantly, they also contributed to the family economy by chopping firewood, picking wild berries, caring for livestock, hunting and trapping, logging, laundering, sewing, canning, preserving, concocting herbal remedies, and much else. Such children may not have learned formal academic subjects in their homes, but they learned skills and social habits that they would take with them into adulthood, especially the values of thrift, hard work, and sobriety that were taught by elders through an oral culture thick with stories and sayings. They also imbibed from family and friends race prejudices and a sense of their own social place relative to the more well-to-do. Finally, they learned how to be parents themselves. Thousands of working-class Americans were reared mostly by their siblings while both parents struggled to keep the family fed. While such an experience taught valuable lessons, many who spent their childhoods parenting smaller siblings looked back with melancholy at childhoods lost.[51]

We have seen in this chapter how, despite the residual formal home instruction that took place on the margins of society, most Americans by the early twentieth century had fully embraced the notion that children should learn in schools. School was fast becoming the defining life experience of nearly all American youth. Before the century ended, however, this consensus was falling apart, so much so that a significant movement emerged attacking the very notion of institutional schooling and attempting a return to home-based education. The next two chapters will describe both why and how homeschooling happened.

Chapter Four

Why Homeschooling Happened, 1945–1990

In 1949 Norbert and Marion Schickel went off the grid. Norbert dropped out of MIT, Marion quit teaching kindergarten, and the couple bought a farm near Ithaca, New York. Strongly influenced by the Catholic Worker movement, the Schickels wanted to get back to the land through subsistence farming. The first few years were very difficult. Norbert spent his days reading anything he could get his hands on from Cornell University's agricultural extension program and talking with neighbors who knew something about farming. The children started coming fast—thirteen in all. Very quickly they were drafted to help with farm chores. At first the family tried to grow everything they needed: large fields of fruit trees, a huge vegetable garden, and livestock provided year-round food for the growing family. But despite their best efforts it soon became clear that supplementary income would be necessary. After a few years of paltry crops, the Schickels shifted their energies to dairy farming, and the children quickly became adept at milking and selling. In 1953 the family purchased the adjoining farm, giving them 225 acres for their growing herd of cattle. At its peak the family farm had sixty-five head.

When the time came for the eldest daughter to go off to school, the Schickels decided they would try teaching her at home on the farm. Marion herself had spent four years of her childhood in the 1920s being tutored in the home of a widow, who supported her family in this fashion. Marion was fairly confident that three years at Wheelock College and private school teaching experience had prepared her well enough to do the job. In between chores Marion and Norbert together taught all of their children to read and do basic math inside the farmhouse. There were flashcards, Cuisenaire rods, a huge painted blackboard covering one wall, maps from National Geographic all over the place, and books donated by a sister-in-law who was a nun. Norbert would read to the children during lunch and dinner and Marion would do the same before bed—*Huckleberry Finn*, the *Little House* books, biographies, and historical works. As the children grew and word got out of Marion's skills, neighbors began sending their children, some also offering to help teach certain subjects. One neighbor taught the girls to sew. Another provided music lessons. One father would come to the house and teach hands-on science classes. Anything and everything became fair game as an intellectual or moral lesson. Cooking, furniture-making, bookkeeping for the farm—all of these taught the children math and more intangible things as well. On every New Year's Day Norbert would reveal the family "word of the year," a moral precept that would be everyone's focus of attention: trustworthiness, responsibility, fairness, kindness.

When the Schickels first began their home school, the local superintendent of schools called to make sure the children were receiving adequate instruction. For the first three years Mrs. Schickel submitted curriculum plans to the school district, naming her efforts the "Mary Hill Country School." Impressed, the school district not only accepted her work but also actively sought her out to deal with some of their problem cases. Since the local public school had little by way of special education, on several occasions the principal would recommend to parents of children with special needs that they visit Mrs. Schickel. She later recalled one child in particular who had been dubbed "the worst-behaved kid in the school" and who, though in fourth grade, could hardly read or do basic math. A year of individualized instruction and attention mellowed the boy considerably and got him up to speed in reading. He went on to become a successful engineer. Another child with more severe physical problems was sent by the school district to Mrs. Schickel, who patiently worked with her until she was ready to enter a boarding school and eventually to lead an independent life. Even children who didn't go to school with Mrs. Schickel would often stop by the farm after school to play with her kids and sometimes do farm chores. One barn became a full-court basketball arena. With the addition of a backstop, a pasture became a football and baseball field. Given such rich childhoods, it is not surprising that all thirteen of the Schickel children went on to college and five of them received advanced degrees.[1]

In the 1950s the Schickel family was a true anomaly. By the 1980s, however, there were thousands of families all over the country doing what the Schickels did. What happened? The next four chapters will attempt to answer this question. In this chapter I will explain large-scale social trends that made homeschooling, if not inevitable, at least a plausible option for a diverse group of Americans. In ensuing chapters I will examine more precisely how homeschooling took on the form it did by looking into the leading activists and organizations that created the movement and the legal and legislative history of the homeschooling question.

Suburbia and Its Discontents

As we saw in chapter three, by the 1930s nearly all American children were experiencing at least some formal schooling. The Depression played an important role in accelerating the trend. With jobs scarce even for adult men, children had little better to do than go to school. From the 1930s onward the country saw steadily rising rates of school attendance, especially in the higher grades. In 1930, just under half the children between ages fourteen and sixteen were in school. By 1950 there were over 77 percent in school. Growth in all sectors of schooling continued unabated. The school year lengthened (from 144 days in 1900 to 178 by 1950). Local districts consolidated into larger organizational units (from 117,000 districts in 1939 to 41,000 in 1959). Textbooks and buildings became more standardized. National tests, professional organizations, and federal involvement further homogenized schools across the country. By 1970 sixty million Americans were enrolled in some sort of school, and 80 percent of school-age Americans were graduating from high school. This profound expansion and standardization is the fundamental fact without which the homeschooling phenomenon makes no sense. Homeschooling, like so many of the other significant cultural movements of the 1960s and 1970s, was very largely a reaction against the mass culture of the modern liberal state, a culture realized perhaps most perfectly in the consolidated public schools located on metropolitan outskirts amidst the rapidly expanding suburbs.[2]

The Schickels were bucking the prevailing trends when they moved to the country in 1949. The family farm had been in decline for decades by then, killed off by mechanization. The invention of the combustion engine led to the tractor and other labor-saving devices that transformed agriculture. In 1900, of the total population, 32 percent still farmed. By 1940 only 23 percent did. By 1980 just 3 percent of Americans still lived on farms. The automobile eviscerated the city as well. First of all it made city streets less pedestrian-friendly and far more congested. More profoundly, it gave people a way out. Henry Ford, architect of the revolution, predicted, "the city is doomed. We shall solve the city problem by leaving the city."[3]

We did leave the city. Earlier suburbanization had occurred along rail lines, with settlements extending in thin tendrils out to the countryside. But the automobile opened up vast new tracts of land for development, allowing homes to be built far from places serviced by public transport. By 1915, two and a half million Americans owned cars. By 1925, twenty million did. By 1955, sixty-three million Americans owned automobiles. All these drivers of course needed places to go, so road construction followed swiftly. In the 1920s a coalition of pressure groups—tire and auto parts manufacturers and suppliers, oil companies, service station owners, road builders, and land developers—successfully petitioned local governments to finance roads not by tolls but by general taxation. By 1929 every state in the Union had a gas tax to underwrite its road building. The "road gang" also succeeded in getting the federal government involved: the 1921 Federal Road Act provided matching federal funds for important local roads and began planning an interstate highway system; the 1944 Federal Highway Act raised the federal contribution to 60 percent and expanded road eligibility for funding; and in 1956 President Eisenhower signed the Interstate Highway Act, ostensibly as a national defense measure, but in actuality a massive pork barrel project providing 90 percent of interstate highway money but giving states freedom to design and locate them as they pleased. This infrastructure quite literally paved the way for mass suburbanization. On average, 883,000 new homes were built per year between 1922 and 1929, most of them on the outskirts of cities. After a lull during the Depression and World War II, development picked up again on an even more massive scale. There were 937,000 new homes built in 1946. By 1950 over 1.5 million new homes were being built a year. By 1980 over 40 percent of the population, more than 100 million people, lived in suburbia.[4]

Government subsidies did not stop with road construction. The Federal Housing Administration (FHA) and the Veteran's Administration (created in 1944) provided mortgage guarantees on new home purchases, allowing developers to build and sell at will, confident that the Federal Government would bail out anyone who couldn't pay. Developers also received huge tax breaks for such commercial projects as strip malls, fast food restaurants, industrial parks, and gas stations. Government underwrote sewerage, zoned undesirable public housing away from suburbs, gave developers virtual free reign over land use, located Department of Defense sites in outlying areas, and drafted income tax laws allowing deductions for mortgage interest and property taxes. As one historian has put it, "Sprawl became the national housing policy." Given such extravagant government largesse, it is little wonder that few middle-class Americans during the 1950s and early 1960s expressed concerns about "tax and spend" liberal programs, and it makes the subsequent libertarianism that has typified so much of later suburban politics more than a bit ironic.[5]

There were two main reasons government was so committed to suburbanization. The first relates to the cold war. Suburban sprawl decentralized the population, making it more likely to survive an atomic attack. More importantly, it facilitated the "ownership society" that would quell domestic labor unrest by giving workers a slice of the domestic dream and a platform for consumption. Abraham Levitt, the most influential of large-scale suburban developers, claimed that "no man who owns a house and lot can become a communist." Mr. Homeowner and Mrs. Consumer seemed eager to prove him right. Each home became a shrine to products: lawn mowers and barbecues, furnishings and draperies, swing sets and televisions. In 1959, Vice President Richard Nixon used the modern "state of the art" kitchen to demonstrate the superiority of private enterprise to Soviet Premier Nikita Khrushchev at the Moscow Trade Fair. If collectivism was the enemy, the suburban home with its private back yard, garage, and, by the 1960s, air conditioning, was the picture of individualism. Cities might have concert halls, opera houses, ballet companies, museums, and other communal forms of entertainments, but suburbia had private living rooms with televisions. By 1976 the average American watched 28 hours of TV a week, where suburban values were modeled in sitcoms and reinforced via an endless stream of commercials.[6]

The second reason for government interest in suburbia pertains to race. In the 1940s and 1950s government policies facilitated the separation of white and black Americans in many ways. Federal highway money was used by local interests to build "white men's roads through black men's bedrooms," constructing physical barriers between black and white neighborhoods. Many neighborhoods with majority black populations were deemed too risky to be insured by the FHA. Decisions by developers such as Levitt, who publicly and officially refused to sell to blacks for two decades after World War II, were legal and generally approved of by many government agencies. And then came *Brown v. Board of Education* in 1954, which declared segregated schools inherently unequal and therefore unconstitutional. Public schools had already been a powerful draw for many suburbanites. After *Brown*, they became crucial, for not only did they offer newer and generally better facilities than urban public schools, but they also provided millions of white Americans with a way to sidestep desegregation. The Supreme Court's *Swan v. Charlotte-Mecklenburg* decision (1971) that led to court-ordered busing within school districts accelerated the flight of white Americans from cities to suburbs, whose self-contained districts freed them from having to integrate. With 1974's *Milliken v. Bradley* decision, the court (in a five to four ruling) declared that cross-district busing was not required, and the *de facto* segregation of whites in suburban districts and minorities in urban districts was cemented. From that time to the present, many parts of the country have become more segregated than they were before *Brown*. This remains the case despite the Fair Housing Act of 1968 that swept away the legal basis for segregation in housing. Since 1960, over twelve million African Americans have also moved to the suburbs, but the great majority of them have tended to settle in predominantly black neighborhoods. Segregation by race thus remains the norm, though it is technically illegal. Segregation by class is even more pervasive and legally enforced through zoning regulations concerning lot size, single-family dwelling requirements, and many other markers of wealth.[7]

This mass movement toward suburban life impacted American families in many ways. Much has been written and said about the fate of suburban women, a good bit of it in response to Betty Friedan's famous *Feminine Mystique*, first published in 1963. In that work, Friedan argued that suburban life served as a sort of comfortable concentration camp for women, segregating them inside the walls of domestic bliss from adult conversation, meaningful work, and political involvement. Many women truly did feel much of what Friedan was describing. One suburban housewife reported the following on a questionnaire in the late 1950s:

> Because of the size of our family, we have very little personal fun—I mean no clubs or activities. I used to be very active in PTA, church (taught Sunday School), and garden club, but my last two children, now 4 and 2 years old changed all this. I just stay home with them and taxi my oldest boys around....I feel quite stale as though I don't use my mind enough.

Friedan's critique joined a host of other works of the 1960s and later years, which have consistently portrayed postwar suburbia as a "smug and phony world." However, more recent historians have uncovered a very different 1950s, arguing that the stereotype reflects more the dystopian vision of succeeding decades than historical reality. While many women did express concern over "cultural isolation," especially given their separation from extended family, postwar suburban women on the whole were far more engaged civically than the stereotype would allow. The suburban home was often the springboard for aggressive political involvement. Women organized locally to fight smut, to promote or hinder integration, to defeat communism, to add a traffic light here or change zoning laws there. They were particularly enervated by school-related issues. In many respects conflicts between parents and school officials in the 1950s "set the stage for the residents' negative reaction toward the integration plans" that were to come later. Many postwar suburban women began to feel a "growing disenchantment with the state" even before the events that we think of as "the 1960s" happened.[8]

Women's roles were changing too. The most popular television figure of the decade was not June Cleaver but Lucy Ricardo, a woman constantly (and hilariously) navigating between public aspirations and domestic duties. Movies and print culture depicted the 1950s family as far more complex than commonly supposed. Joanne Meyerowitz' exhaustive study of popular magazines during the period concludes that the feminine ideal was not the happy housewife but the woman who could achieve public recognition without sacrificing her femininity and domesticity. The reason for all this stress on the public woman is clear. Women's employment outside the home had been increasing slowly in the decades before World War II. After the war more and more women, especially married women, went to work. Each decade between 1940 and 1990 saw a 10 percent increase in the percentage of married women in the work force. By 1960, three times as many wives were working than had been in 1940. By 1985, of women with children under six, 50 percent were in the work force. Increasingly, it was middle-class mothers who were making this choice, driven largely by a desire for more consumer goods. Many of these women struggled to reconcile their newfound public roles with undiminished domestic duties and looked for role models and advice anywhere they could get it.[9]

Family changes also impacted men and children. In many ways the 1950s saw the mainstreaming of the "companionate" ideal of marriage and child rearing popular among the upper classes earlier in the century. Marriage was to be an emotionally fulfilling experience, with both husband and wife committed to loving friendship, a satisfying sex life, and sacrifice for the kids. In 1954, *Life* magazine announced "the domestication of the American male," whose career was simply a means of providing for the family's insatiable demand for consumer goods. Fathers were advised and expected to help out at home more than previous generations of dads had done, and the evidence suggests that many obliged. What little remained of patriarchy was increasingly replaced by what one historian calls the "filiarchical" family preoccupied with "making children happy." The suburban middle-class family routine became organized around ball games, school schedules, dance classes, and a host of other activities, all suggesting to the children "that they were the center of the universe."[10]

For parents brought up to respect authority, schooled longer than any previous generation in American history, come of age during economic crisis, matured by war, and now flush with the prosperity of American technological progress, it made sense to trust what the experts were saying about family as in everything else. No expert captured the national mood better than Dr. Benjamin Spock, whose *Baby and Child Care*, first published in 1946, was the runaway favorite among a large batch of postwar child-rearing manuals, nearly all of which sought to replace authoritarian, schedule-driven methods with more gentle, commonsense, child-centered approaches. Though the explicit message was "you know more than you think you do," implicitly Spock and his many imitators were inculcating the notion that the experts know better and should be consulted on every topic, and parents were quick to do so. Mothers especially had an insatiable appetite for expert advice, though historians studying their responses to Spock and other authorities have found a consistent pattern of selective appropriation. Experts might often sound condescending ("when the school provides materials judiciously chosen, and parents are too wise to intrude, the combination is a winning one" noted one educator) and sometimes inane ("If he is not yet skilled at dressing himself," one parent educator advised, "he can practice every day"), but most parents accepted without question the abiding message that a happy, healthy home would result in secure, well-adjusted kids.[11]

While the message was one of security and life adjustment, anxiety and self-doubt pervaded both parents and the experts. Many parents felt unequal to the task of raising successful children. Having been schooled by experts to think that psychological maladjustment in their children was the result of bad parenting techniques, many parents worried that they had "failed to do everything necessary to make contented, cooperative human beings of our offspring." Experts, while sounding authoritative in advice columns and parenting manuals, expressed anxieties of their own when they talked together.

The White House Conference on Children and Youth in 1950, whose chosen subjects were "the healthy personality" and "personality in the making," revealed a deep sense of self-doubt among the luminaries making up the program. They worried that science had discovered very little about child development and environmental influences, that most of their views were based more on theory than empirical data. Throughout the 1950s, intellectuals and other public figures grew increasingly troubled by the mass society their own expertise was helping to construct. Bestselling books such as David Reisman's *The Lonely Crowd*, William H. Whyte's *Organization Man*, John Kenneth Galbraith's *Affluent Society*, and Vance Packard's *Hidden Persuaders* consistently sounded the alarm that bland conformity was draining the color out of American life, leading to alienation and malaise. We are accustomed to think of the 1960s as a rebellion against the staid conformity of the 1950s. But the popularity of these and other books praising individualism suggests, in the words of Thomas Frank, that "the meaning of 'the sixties' cannot be considered apart from the enthusiasm of ordinary, suburban Americans for cultural revolution."[12]

When the cultural revolution did come, many Americans were taken by surprise. Many postwar parents were at heart individualists despite the conformity of their outer lives. Mothers tended to dominate at home despite a formal commitment to father's authority. Both parents raised their children by Spockean methods that encouraged individuality and self-expression. But when their children "set out to complete the half-kept promise of their parents" by challenging core cultural standards, parents didn't know what to do. Some blamed Spock for creating, in Norman Vincent Peale's words, "the most undisciplined 'generation' in history," lamenting that kids had been "Spocked when they should have been spanked." When Spock was arrested in 1967 for peace protests, there was no national outcry on his behalf. His activism only confirmed what many feared, that permissive child-rearing techniques had led to the collapse of "respect for authority, for the school, for the family." But though Spock the man was scapegoated, his ideas continued to dominate popular thinking about child rearing. Even the most conservative of Americans would agree with his basic belief that "care of children and home is at least as important and soul satisfying as any other activity," and that no one "need to apologize for deciding to make that their main career." The key to successful child rearing for Spock was always "a mother's love." In meeting the needs of her children, "she, in turn, would find fulfillment."[13]

After Spock, advice manuals grew more strident and partisan, with one wing supporting a return to discipline and the other embracing a more child-centered approach. But underneath the rhetorical war between conservatives like James Dobson, Dan Kindlon, John Rosemund, and Gary Ezzo and liberals like Barry Brazelton, William Damon, Stanley Greenspan, and Penelope Leach, both sides advocated a home culture of "structured commitment, cooperation, and communication" that was vintage Spock. Like Spock, both sides wanted to steer a middle course between "authoritarian" parenting, which all agreed was overbearing, and "permissive" parenting, which was too indulgent.[14]

This is not to say that the polemics were not important. They were, for two contradictory reasons. First, the strident tone garnered publicity and sold lots of books. The battle over parenting style became a front in the larger culture war of the 1980s and 1990s. By 1997 five times as many parenting books were being published annually as had been in 1975. But at the same time, the increasingly strident tone and partisanship contributed to a widespread rejection of expert authority in general on these matters. By the 1970s a clear rebellion against expertise was fomenting among many American parents. One father spoke the 1970s' mood when he proclaimed, "we parents have to start believing in ourselves again and boot out all the experts who tell us they know what's best for our kids." At the White House Conference on the Family convened by President Jimmy Carter in 1980, the people themselves were the experts. Unlike previous White House efforts, this conference attempted to create federal policy not by convening the best expertise on the question but by listening to parents. The original plan was for three regional conferences where "representative Americans" would be placed in focus groups and asked to brainstorm a national agenda. The results, however, were calamitous. Very quickly the first conference, held in Baltimore, devolved into a shouting match between feminists and right-wing activists. Conservatives walked out in protest before the final vote, arguing that the proceedings, in Phyllis Schlafly's words, were "contaminated by the liberal Carter machine." The walkout led to the passage of fifty-seven proposals reflecting a decidedly liberal orientation. At the next conference in Minneapolis the conservatives stayed until the end, but the final product was an incoherent hodgepodge from which it would have been nearly impossible to build federal policy: Conservatives passed resolutions condemning homosexuality and secular humanism. Feminists won affirmations of the Equal Rights Amendment and Abortion rights. Bottom-up policymaking pleased nobody. And

Ronald Reagan's ascent to the Presidency made the entire question moot anyway.[15]

The White House Conference, despite its failure, illustrates many important trends in American family life at the end of the twentieth century. First of all, it was a response to real family problems. Births to unwed teens increased by 38 percent in the 1970s. The divorce rate doubled between 1966 and 1976, abetted by the new "no fault" divorce laws pioneered by Ronald Reagan, himself divorced, in California. While premarital intercourse was rising sharply, the birth rate was plummeting. By 1980 even modernists were worried about the American family. Margaret Mead asked, "Can the family survive?" Psychologist Urie Bronfenbrenner worried that "America's families are in trouble—trouble so deep and pervasive as to threaten the future of our nation." Others noted that children were spending more time watching TV than they were in school or with parents, exposing them to "a commercialized, sexualized, violent media culture" which led to premature "adultification." Apocalyptic bestsellers like Hal Lindsey's *Late Great Planet Earth* couldn't be topped for their cosmic pessimism, but writers on children and the family, both conservative and liberal, gave it their best effort.[16]

The White House Conference's rejection of expertise also reflected growing trends. Missouri Congressman William Hungate succinctly summarized the shift in public opinion about political leadership, "Politics has gone from the age of 'Camelot' when all things are possible to the age of 'Watergate' when all things are suspect." Disillusionment with government extended to all sectors, including schooling. Parents looked on as fights between teachers and administration got nasty. They worried about the records schools kept on their children and wouldn't let them see. Some conservative parents protested against schoolbooks that mentioned witchcraft, evolution, world government, pacifism, and other cultural flashpoints. Sex education, life adjustment, progressive pedagogies like the "new math" and whole language reading instruction came under attack. More court cases were brought against public education between 1969 and 1978 than there had been for the previous fifty years. Schools took it from the right for being insufficiently intellectual, in titles like *Educational Wastelands*, *American Education: A National Failure*, and *The Literacy Hoax*. Left-leaning books indicting the authoritarianism of public education were even more merciless, bearing titles like *Growing Up Absurd*, *Death at an Early Age*, and *Crisis in the Classroom*. Finally, court-ordered busing was, for many, the last straw, making "bitter and immediate antagonists of parents" in many parts of the country.[17]

This growing animus against government schools, however, coincided with an ever increasing reliance on them by a growing number of people. Furthermore, the growth of government oversight of families did not stop with schooling. Americans seemed both to fear Big Brother and to accept the need to protect children from "the usually private activity of child abuse" and other threats to their well-being. Adult concern for children's safety bordered on the hysterical as national media reported case after shocking case of child murder, pedophilia, cultic activity among children, and drug lords preying on kids. Panic over child safety led to calls for government action that tended to creep toward government oversight of adults as well. Despite growing antigovernment sentiment, new government initiatives were everywhere: health crusades such as inoculations and diet and exercise regimes, welfare policies, antidiscrimination initiatives, laws making it illegal to deny birth control to unmarried women, tax breaks for institutionalized childcare and other incentives for working mothers, and of course the 1973 *Roe v. Wade* decision legalizing abortion throughout the country. The White House Conference of 1980 reflected both this growing federal involvement in the most intimate aspects of private life and the growing resentment such oversight was engendering among many Americans.[18]

Privacy and the Homeschooling Option

Without question, one of the most significant developments in recent American social life is the fragmentation of much of the population into two factions. Call it what you will: conservative and liberal, right and left, red and blue. Since the 1980s, commentators have been much exercised over the division of the country into warring camps on most social issues. But what is often missed in such an analysis is the underlying symmetry of vision both camps tend to possess. The cultural left and right may argue incessantly, but they speak the same language and share a similar set of background beliefs. Since the 1960s, Americans on both sides of the political spectrum have been more interested in local community and self-determination than in national identity. Historian David Farber has shown how "calls for a more direct democracy built on local control and community right to self-determination" came in the 1960s "from Black Power activists, Chicano militants, white Southerners, and white urban ethnic blocs." Conservative and liberal Americans had radically different private visions of the good life, but they all shared a commitment to private vision. Private health clubs and private schools emerged to replace public recreation and education among all sorts—civil rights activists and white segregationists alike. In the early 1980s California conservatives closed public restrooms and slashed funding for public libraries and parks. Massachusetts liberals privatized snow removal and garbage collection at the same time. In 1970 there were 7,000 homeowners associations making private rules for private communities. By 1980 there were 60,000. Conservatives are often seen as the advocates of free-market capitalism and limited government. But capitalism worked for radicals like Jane Fonda and Jerry Rubin too, both of whom made fortunes as entrepreneurs. Entrepreneurial initiative was championed by Sunbelt conservatives and black power advocates alike. Everyone watched *Roots* (130 million viewers), a show symbolizing the rejection of melting-pot forgetfulness and celebrating particularity. Everyone went casual: blue jeans and t-shirts ceased being a badge of outsider status and symbolized instead an embrace of the informal, the authentic. "Conservative" churches were anything but conservative in their celebration of private, direct experience of God and their appropriation of countercultural music and hairstyles. And everybody waxed apocalyptic, whether they be Christians discerning the antichrist's immanent arrival in the latest headlines, or hippies predicting an environmental holocaust. Both groups saw themselves as the small faithful remnant surrounded on all sides by the forces of darkness. By the 1980s young Americans on both the left and the right had largely given up on building a better America, hoping instead to "build alternative institutions and create alternative families—a separate, authentic, parallel universe."[19]

Given this pan-ideological commitment to local, authentic, private life and contempt for establishment liberalism, it is not surprising that members of both the countercultural right and the countercultural left began to practice and advocate homeschooling. In the following discussion, I will look at each side separately and then draw some conclusions as to why homeschooling proved attractive to both.

First for the left. We could start the story perhaps with the 1968 Democratic National Convention in Chicago. Thousands of young radicals had flocked there to protest the Vietnam War, and Mayor Richard Daly had commissioned 25,000 law enforcement officers to control the crowds. For a week the two sides coexisted in a tense but stable situation. But on Nomination Day, 15,000 protesters moved into downtown Grant Park for a rally sponsored by the National Mobilization Committee to End the War in Vietnam. For some reason the police suddenly snapped and began indiscriminately beating, clubbing, gassing, and arresting protesters, many of whom fought back as the battle moved toward Michigan Avenue. The nation watched in shock as their televisions broadcast vivid images of police brutality and student revolt. Similar scenes were played out at colleges and universities around the country, most notably at Kent State University where four protestors were killed by National Guardsmen. Many a peace activist recoiled in horror from a revolution that had suddenly turned violent.[20]

By 1970 a sense of despair and powerlessness had gripped many leftist activists. Years of protest had failed to stop the Vietnam War. Many came to doubt the very possibility of a political solution, concluding that any real change must come instead from a compelling alternative society. Student protests of the late 1960s with their grand public visions gave way in the 1970s to what Marianne DeKoven has called "utopia limited," as radicals turned toward small-scale communities. Many went "back to the land," and many formed communes, committed, like the Pilgrims of old, to modeling a new society in hopes that the old world somehow would be won over.[21]

In contrast to the synthetic fabrics, day glo colors, and plastic dreams of the 1960s protest movements, the cultural left in the 1970s was earth-toned and organic. "We're learning self-sufficiency and rediscovering old technologies that are not destructive to themselves and the land," one communard explained. By the early 1970s there were some 2000 rural communes in existence and perhaps as many as 5000 less organized "collectives," ranging from urban villages to more informal "crash pads" where anyone was welcome to a couch or space on the floor. Some communes were located in abandoned towns, others out in the woods (sometimes on public land). Most were small—one or two dozen people. But there were enough of them to sponsor an underground economy mediated by the *Whole Earth* catalog, which taught readers how to build a house, raise animals, and grow crops even as it sold them the products they would need to do these tasks. They were inspired and instructed as well by *Mother Earth News*, which enjoyed a circulation of 400,000 at its peak in 1978. Though the rhetoric was profoundly countercultural, the agrarian and do-it-yourself spirit pervading this movement was classic populist Americana.[22]

Family was a major concern of many communes. Several even had the term in their name: the California "Lynch Family," the New Mexico "Chosen Family," the New York "Family." Given the commitments of so many of these groups to countercultural lifestyles, there was quite a bit of experimentation with family style. Some communes practiced open marriage, where all belonged to all, or group marriage, a more limited but still far from monogamous situation. Some tried alternating partners on a regular schedule. One commune had three males mate with a woman during her fertile time so that no sense of private ownership of a child would emerge. But very quickly close communal living and the hard work of staying housed and fed put a damper on sexual license and experimentation. Most colonies that lasted became more or less monogamous in practice even if they remained open to innovation in theory.[23]

With coupling came childbirth. Many communes saw natural childbirth as the quintessential statement of their philosophy of getting back to nature and dispensing with the military-industrial complex. Some developed elaborate rituals and celebrations around the birth event. More than a few communes were landscaped with trees planted in honor of a new child, fertilized by the afterbirth. Communes became nurseries for a new wave of midwifery as necessity bred expertise. The modern day home-birth movement owes a good deal of its success to the work of some of these communes, perhaps most notably The Farm in Tennessee. Once born, commune children were typically parented very permissively, according to a philosophy often called "attachment parenting." One research team studying several communes noted that children "almost never leave their mother," being breastfed as often and as long as they desired it. But as the children got older, complications surfaced. Many of the adults in communes were deeply committed to personal autonomy and against external rules and prohibitions. As one researcher noted, "many hippies, including communal mothers, tend to regard their lives as unsettled, their futures uncertain, and are generally unwilling to sacrifice their own personal questings (for meaning, identity, transcendence, etc.) to full-time devotion to child rearing." One mother put the dilemma well, "What I wanted was a *baby*; but a *kid*, that's something else."[24]

Communes came up with varied approaches to this problem. Many of them believed in group parenting in theory. In practice, however, this often meant very little parenting at all. For many, children were merely accepted as miniature adults and given full entry into the life of the commune. Thus they learned by imitation, participating in folk art and music, gardening and husbandry, food preparation, and of course drugs. Drugs formed the organizational basis for quite a few communes. Marijuana was prized for its association with the "peaceful, easy feeling" and LSD for its promotion of powerful spiritual experiences that led many to a sense of clarity about life's simplicity.

Researchers found that children "were exposed to drug experiences at an early age. The feeling was not that they should be deliberately given drugs but if they expressed interest or curiosity in having a drug experience then they were allowed to participate." Though the license given to children in many communes startled researchers, they consistently noted positive outcomes from this "inadvertent" sort of education. One group found that the teenaged children they witnessed seemed not to experience anything like "adolescence." Others noted the consistent maturity, self-confidence, ease around adults, and independent spirit of the youth they met. One research team, a married couple with children of their own, concluded after several months of studying many communes:

> While we began the book with the suspicion that a hippie child is a wild child, we ended up believing that well-behaved children are the most radical alternative to American society. The farther away from regular families and cities and careers that we get, the less obnoxious and self-centered the kids get.

Though the pair did not like the communes they visited, they were amazed at the positive change in their own children that resulted from the exposure to alternative living. "It is a little like seeing the miracle and then turning down the religion," they said.[25]

Some communes were more deliberate in their approach to education. Many began with a philosophical antagonism to public education. One communard expressed it clearly, "suddenly I saw all the bulls—t in the whole educational and social system.... The problem with our schools is that they are turning out robots to keep the social system going." Many viewed schools as the primary means of assimilating children into "the establishment," their mortal foe. Larger communes often started their own schools on the premises, and many turned to homeschooling. As Herb Goldstein of the Downhill Farm in Pennsylvania noted, "Some wanted to homeschool because they wanted their kids to do better than they could if they went to public school." Stacia Dunham's childhood illustrates what this better education might look like. Her father had sold his Southern California jewelry business in 1974 to take his family back to the land. They bought an old farmhouse near some like-minded souls and set out homesteading. Stacia shared a bedroom with her five sisters. The family spent most of its time staying warm and fed. "I remember one year we caught five hundred trout" she recalled, "We'd sit at this huge table, assembly-line fashion, cleaning and gutting fish." The children also cared for goats, chickens, pigs, and a cow. They tended a large apple orchard, foraged for berries, hunted and dressed the kill, and even panned for gold. "We'd actually find little nuggets sometimes. My father had been a jeweler, so he would take them and sell them to his friends." Evenings were spent reading together. Math was learned when necessity required it, project-method style.[26]

The results of communal and back-to-the-land education were mixed. Most adults who grew up in communes look back with fond memories. Many children so raised went on to get advanced degrees and have successful careers. One mother who had raised her children in a commune described the results like this, "the kids turned out to be bright, creative, interesting and full of life. It's almost as if being exposed to all the wildness back then demystified that way of life for them." Her own daughter became a doctor and her son a carpenter. Very few chose to live in communes themselves. One long-time communard explained why, saying that back-to-the-land kids "are the ones that have grown up and turned into Young Republicans. They weren't about to do what their parents did." Surveys of adults who were raised in communes have found that the only regrets most have were the names given to them. Many a young Vishnu or Ongo Ishi changed their names to Bill and Samantha as soon as they were able. Some also regretted having had so few peers to grow up with. But most seemed to survive the early exposure to drugs and sex with little trouble.[27]

Some "graduates" of communes, however, were not so sanguine about the experience. Some women especially look back with a profound sense of sadness at the chaos of their early lives. Moon Zappa recalled,

> At my house there was no supervision, so there was no reason to sneak. At my house there were no rules, so there was nothing to rebel against. I hated it.... I craved rituals and rules like my friends had. I prayed for curfews and strictly enforced dinner times. Uniforms and organized events and people with *goals* amazed me.

Elizabeth She' describes a childhood brutalized by the ideology of free love, "If you ask me, free love ain't either. It's not love, and it's not free. I've been paying the price for thirty years." Girls seemed especially vulnerable to abuse by adult males in an environment where there were "no boundaries, no guidance, no protection. Nothing was sacred." Some express deep regret over a childhood marred by broken relationships, fatherlessness, drugged out and distracted adults, grime, and poverty. Others, such as Rain Grimes, explain how a countercultural girlhood could lead to a different sort of rebellion, "My friends went to college and became vegetarians. I went to college and became a meat eater.... I lust after processed sugar, red meat, full-fat dairy. It is the legacy of growing up vegetarian and sugar-free."[28]

Though there are some standout exceptions, most hippie communes did not survive the 1980s. Well-meaning idealists who sought to create a limited utopia free of the rules and regulations of mainstream society attracted to their experiments all sorts of drifters, lechers, and opportunists who took advantage of their naiveté and destroyed their communities. Many communes that did have rules still disbanded due to personality conflicts among members, a situation almost impossible to avoid given the close quarters and limited resources most communes possessed. Independent subsistence farmers fared no better. Eleanor Agnew generalizes the life course of many of them:

> A person goes to the land to be self-sufficient and free, the freedom loses its luster when the poverty grinds, the person and his or her spouse divorce, and the person slides back into the mainstream, gets a professional job or entrepreneurial gig, and remarries.[29]

While most of their experiments failed, the hippie wing of the antimainstream movement continues to have a powerful impact on American culture. Since the 1970s, organic food, whole grains, aversion to processed and packaged products, and concern for the environment have entered the mainstream. Natural childbirth and home birthing continue to grow in popularity. Drug use, despite vigorous government efforts to stop it, continues. The left-wing critique of public education and preference for a freer, more natural childhood centered in the home continues, as we shall see, to reverberate in the modern homeschooling movement. Most profoundly, the counterculture's revulsion against conformity and longing for individual expression and authenticity has become the most basic trope of popular culture. We are swimming in advertising slogans like "obey your thirst" and "have it your way." The theme is a perennial favorite of pop hits from Madonna's "Express Yourself" to Natasha Bedingfield's "Unwritten." It dominates the moral vision of the cult of self-esteem in educational programming. It is just as popular in religious circles, discernable in Sunday school sing-alongs like "Search all the world over, there's no one like me," evangelist Bill Bright's first spiritual law, "God *loves* you and offers a wonderful *plan* for your life," and the success of prosperity preachers such as Joel Osteen. The countercultural quest for personal fulfillment and individual self-expression is now as mainstream as it can get.[30]

The Countercultural Right

The New Left "hippie" movement was at heart a religious one. LSD trips very often were experienced and described as religious visions of a society of peace and love. Hippies shared with other religious seekers in the 1970s a preference for ecstatic, direct encounters with God over staid liturgy and boring hymnody. In fact, a large number of the "Jesus people" of the late 1960s and 1970s were converts from the countercultural left. In 1967 Berkeley evangelists with the Campus Crusade for Christ organization went native—long hair, tie-dye, hip jargon. They changed their name to the "Christian World Liberation Front," opened crash pads, and preached that Jesus was a better trip than LSD. Thousands of hippies took to the message and exchanged drugs for Jesus. Christian folk rocker Larry Norman, a poster boy of the movement, captured the sentiment with his memorable lyric:

> Gonorrhea on Valentine's Day,
> And you're still looking for the perfect lay.
> You think rock and roll will set you free,
> You'll be dead before you're thirty-three.
> Shooting junk till you're half insane,
> Broken needle in your purple vein,
> Why don't you look into Jesus, he's got the answer.[31]

But as hippies became Jesus Freaks, it was not just their own lives that were changed. American Protestantism itself was transformed in the 1960s and 1970s. Two basic shifts occurred. First, the old denominational distinctions (Methodist, Baptist, Episcopalian, Presbyterian, etc.) began to fade, replaced by a sharp binary between "conservative" and "liberal" churches. I put the terms in scare quotes because churches that self identified as conservative were usually quick to throw out traditional forms of worship in exchange for more casual and emotional styles, and those deemed liberal often held on to these forms. But "liberal" churches, even if they still worshipped in a manner consonant with earlier centuries, tended to shy away from the more miraculous and exclusive claims of Christianity—Virgin births, miraculous healings, resurrections—such things were seen as the embarrassing relics of an unenlightened age. In contrast, "conservative" churches celebrated such things, completely embracing the supernatural elements of the Bible. While liberals preached a gospel of ambiguity and unanswerable questions, conservative religion offered a real God who could be known personally and whose Book offered short, simple answers to everything: "Jesus is the answer for the world today," sang Andrae Crouch. American Protestants realigned themselves according to this divide, and the results were good news for conservatives and bad news for liberals. Between 1965 and 1975 alone, the Episcopal Church lost 17 percent of its members, the United Presbyterians 12 percent, the United Churches of Christ 12 percent, and the United Methodists 10 percent. In contrast, Southern Baptists increased by 18 percent during the same span. Independent Bible-based churches and especially Pentecostal groups flourished, joining a fierce biblical literalism to a modern, emotional worship style. The Assemblies of God, for example, grew by 95 percent between 1973 and 1987. Dramatic growth in the conservative, separatist sector spawned a host of alternative cultural institutions that mimicked even as they condemned the cultural mainstream: Christian bookstores (1400 new stores between 1971 and 1978), romance fiction, radio and television stations, rock concerts and festivals, music awards, theme parks, summer camps. A parallel Christian culture was emerging that allowed "kids to be normal, blue-jean-wearing, music-loving American teenagers without abandoning the faith,...to be devout without being nerdy."[32]

The second shift came in terms of political engagement. Early in the 1960s studies of voting patterns consistently found that religiously conservative people were the least likely Americans to be involved in politics. Back in the 1960s Jerry Falwell was still preaching that "we need to get off the streets and into the pulpits and prayer rooms." As late as 1971 sociologist Robert Wuthnow explained that conservative Protestants had a "miraculous view of social reform: that if all men are brought to Christ, social evils will disappear.... Evangelical protestant groups largely ignore social and political efforts for reform." It was the liberals at first who were the activists, especially in campaigns for civil rights and women's liberation. But by the mid 1970s Falwell was organizing "I Love America" rallies in state capitals across the country, calling for a national moral rebirth. His "Moral Majority" shifted focus from the immanent rapture to "the nuts and bolts of voter registration." Why the change?[33]

What happened was an infusion of countercultural sensibility into the most conservative segment of the population. Moral conservatives, shocked and outraged by social change, adopted the techniques of the Left to forward their own agenda. Similar dynamics had been present in moral reform efforts of earlier years, of course, but in all of these cases, from prohibition to the anticommunist crusade of the 1950s, government had been on the side of the conservatives. But now conservatives suddenly found themselves fighting not just immigrant vice, urban decadence, smut peddlers, or communist traitors. The enemy had suddenly become their own government. Conservatives had been growing increasingly alarmed by government expansion since the 1950s. Anticommunism led naturally to a distrust of government generally. Groups such as the John Birch Society and the Minutemen popularized conspiratorial ideas about the extent to which communists had infiltrated the federal government (Birch's founder Robert Welch, Jr. even claimed that President Eisenhower was a closet communist.) The popularity of such sentiments and organizations is remarkable. Public opinion polls in the late 1950s suggested that 5 percent of Americans truly believed that communists had direct control of "the U.S. Government, public education, and the National Council of Churches." By 1964 the John Birch Society had an annual budget of $7 million and a membership of at least 50,000. Australian speaker Dr. Fred Schwarz hosted "Christian Anti-Communist Crusade" youth rallies across the country, attended by tens of thousands of young people inspired by his aggressive advocacy for turning leftist tactics to right-wing goals. The Minutemen, an organization sponsoring anticommunist terrorism by training small cells of activists to use firearms, make bombs, and plan attacks, had about 2,400 cells placed around the country and a membership of ten to fifteen thousand dues-paying members.[34]

For our purposes, what is most remarkable about these movements is their manner of organizing. "Organization is our bag," noted conservative strategist Paul Weyrich, "We preach and teach nothing but organization." All of the right-wing groups were based very largely in the nation's suburbs, especially in the Sunbelt region where thousands of "faux Bubbas" worked steady jobs but flocked to NASCAR events, drove pickups, and listened to country music. Here the term "redneck" was worn "as a badge of honor, a fashion statement, a gesture of resistance against high taxes, liberals, racial integration, women's liberation, and hippies." The homes of this "Southronized" white middle class became the front line of anticommunist attack. As one directive from the central office of the John Birch Society put it, "The battle for saving our Republic could well be won or lost in our living rooms." Tens of thousands of middle-class Americans met regularly in their homes to listen to audio tapes by

Robert Welch, swap conspiracy theories, and strategize about how to defeat the communists through local activism. Though many had left the JBS by the 1970s, they took the training in home-based activism they received as Birchers into other venues, especially the Christian pro-family movement.[35]

Home-based organization meant that women participated in these groups on a massive scale. The new right of the 1950s and 1960s was very largely a women's crusade. As one Pasedena housewife and member of both the John Birch Society and the Republican Women's Club recalled, "women were the core of the conservative movement." Homemakers and mothers did much of the grassroots organizing and not a little of the actual teaching at conservative meetings. These were not Betty Friedan's etiolated domestics. They were empowered, articulate, and unabashedly conventional. As Colleen McDannell has shown, they were the spiritual descendents of nineteenth-century Victorians, trying to preserve a place for domestic Christianity in contemporary society. But though their rhetoric was stridently domestic and antifeminist, their own lives were testimonies to the advances women had been making in education and public life for decades. In the name of the home, these women were coming out of the living room into the public square. They organized reading rooms, voter registration drives, and women's clubs. Many became public speakers, and some ran for office. One activist described herself as "housewife, researcher, lecturer." Rarely did they address this seeming contradiction between their domestic philosophy and public lives head on, but when they did, their words echoed those of first-wave feminists of the late nineteenth century. The president of California's Federation of Republican Women, for example, justified her group's activism like this, "No longer is it possible for [us] to stay home keeping aloof from all outside....Are we through apathy and ignorance going to allow this great dynamic idea we call the U.S.A. to go down the drain of governmental controls and dominance under Socialism?"[36]

No figure better captures the spirit of this female-led, grassroots campaign to save the traditional family from big government than Phyllis Schlafly. Schlafly grew up in St. Louis, attending a conservative Catholic high school with daily Mass and a classical curriculum. She went on to Washington University and then to Radcliffe for her M.A. in political science. In 1945 she took a job as a researcher for the newly formed conservative think-tank, the American Enterprise Institute. Here she became a free-market conservative. In 1946 Schlafly returned to St. Louis, lecturing women on investment strategies and getting involved in local Republican politics. In 1949 she married an attorney and began having children, six in all. She breastfed them and gave them no refined sugar, white bread, soft drinks, or fried food. The family lived on organic produce, raw milk, fertile eggs, and local meat. She taught all of her children at home until the second grade, whereupon they all attended the local Catholic school.[37]

Even while her children were young she became very involved in the Illinois Federation of Republican Women and the Daughters of the American Revolution, helping both groups firm up their commitment to anticommunism by organizing study groups. In 1958 she and her husband formed the Cardinal Mindszenty Foundation (CMF) as a Catholic alternative to Fred Schwarz' very Protestant "Christian Anti-Communist Crusade." Her group had a more international agenda, stressing communist oppression of Catholics in Europe and making the fight with communism not just an economic but a religious battle. By 1961 the CMF was sponsoring over 3,000 study groups in forty-nine states and throughout the western hemisphere.[38]

Schlafly had lost in a run for the House of Representatives in 1952, but her campaign had impressed conservative Republican leaders. She quickly became a popular speaker, especially in the buildup to the Barry Goldwater Presidential campaign. Schlafly's 1964 book *A Choice Not an Echo*, though self-published and grassroots promoted, sold two million copies by the time of the Republican primary and was instrumental in securing Goldwater's nomination. That same year she lost a bitter battle for the presidency of the National Federation of Republican Women. Her more moderate opponents had claimed that a mother of six wouldn't have time to devote to the office. Upon losing, Schlafly withdrew from the organization and began publishing *The Phyllis Schlafly Report*. Conservatives like Schlafly were increasingly isolated from the Republican party after Goldwater's humiliating defeat, and many felt demoralized by leftist gains in the late 1960s and early 1970s. But Schlafly and others got new life in the 1970s when a new cause, the Equal Rights Amendment (ERA), came to her attention. In 1972, when Schlafly began her campaign to stop the ERA, thirty of the necessary thirty-eight states had already ratified it. Feminists at first did not take her movement seriously, but by 1977 the National Organization of Women and ERAmerica had conceded that Schlafly had won over the average homemaker. One deadline and then another passed without the necessary ratifications, and conservatives could claim their first national victory in a long time, a victory that in hindsight was a foretaste of future gains by what would become the powerful religious right. As Donald Critchlow has argued, Schlafly's shift from cold war politics to domestic family issues bridged the gulf between old and new right and paved the way for a resurgent political conservatism united against what it perceived as an elite leftist plot to secularize America by destroying the Christian family. Schlafly and others brought religious conservatives into the political process by turning their attention to issues that would motivate the conservative base.[39]

Without question, the issue that has most powerfully galvanized conservative opposition to big government, especially the federal judiciary, is abortion. Interestingly, conservative Protestants were slow to oppose abortion. Billy Graham was at first mildly supportive of the *Roe v. Wade* decision. The leading evangelical monthly *Christianity Today* put out a special issue that, while expressing various viewpoints, was, on the whole, modestly sanguine about *Roe*. The Southern Baptist Convention regularly passed proabortion resolutions in the early 1970s. In the early years after *Roe*, opposition to abortion rights came mostly from Catholics, and leading the charge was Phyllis Schlafly. Schlafly offered a compelling account of how the ERA, *Roe*, gay rights, and other issues were all cut from the same cloth and must be opposed as a package. Through her advocacy and that of other key Catholic conservatives, many conservative Protestants were won over to the antiabortion cause. By the end of the 1970s, abortion moderate Carl F. H. Henry had lost his job at *Christianity Today* and the magazine was strongly pro-life. But along with this dramatic shift came another, even more significant one. As conservative Protestants and Catholics began to be partners in the abortion war, the centuries-long antagonism between the two Christian traditions began to seem less important. By the end of the 1970s, *Christianity Today* was noting that "Evangelicals are far closer in theology and commitment...to the church of Rome than to many liberals in the Protestant tradition." This was an astonishing historical development. Catholic/Protestant rapprochement on social issues made possible the powerful Christian Right of the 1980s and 1990s. Schlafly, as much as anyone else, had made it happen.[40]

Schlafly's shift to domestic issues also led her into debates about education. Though this aspect of her political activism is less well known, it has been a major thrust of her work of the past twenty years. In 1984 her book *Child Abuse In the Classroom* signaled the new direction and set out her agenda. In 1989 Schlafly began a weekly radio show on education called "Phyllis Schlafly Live" that has run every Saturday since that time, and her Eagle Forum organization sends out a monthly "Education Reporter." Her topics range widely, but homeschooling has been a recurring theme in these programs and in her writings since the late 1980s. Here as elsewhere, Schlafly's views serve as a sort of bellwether of conservative opinion. Schlafly has been so popular for so long among conservatives because she has long had a knack for discerning which issues her clientele care about. And the issue conservative women care about most is education.[41]

At first, homeschooling was not on the agenda of conservative women. In the 1960s most conservative efforts were aimed at keeping public school values consistent with their own. Conservatives had been scrutinizing school textbooks to root out subversively communistic literature since the 1920s. By the 1950s the scene of "neatly dressed, well-mannered women" barging into a school official's office to demand the removal of books deemed too soft on communism had become familiar around the country. In the 1960s the crusade expanded to other issues as well. Conservatives began to rally in opposition to the new focus on social history that paid more attention to the experiences of everyday Americans and less to the founders and presidents. They were particularly upset over discussions of race and sex that tended to make the United States look bad. They despised the "new math" and whole language instruction, which they viewed as pedagogically foolish and potentially dangerous, not only because children trained by these methods couldn't read or figure, but especially because such approaches insinuated that reality is not a fixed given to be learned but an open possibility to be constructed by the individual. Conservatives increasingly worried about sex education, life adjustment curriculum, courses on the topic of death and dying, and readers that included stories about non-Christian religions or occult themes. But while local activism gained them victories in some locales, they rightly discerned that they were losing the battle over control of the nation's public schools. The 1962 and 1963 Supreme Court decisions outlawing organized school prayer and school-sponsored Bible reading shocked and devastated many conservatives. Coming on the heels of the Court's desegregation decisions, many conservative Protestants were simply appalled. Alabama Representative George Andrews spoke for many when he said on national television that the Supreme Court had "put the Negroes in the schools—now they put God out of the schools."[42]

With minorities in and God out, many conservative Protestants left. But even at this stage homeschooling was not really considered. As the Courts pushed to integrate public schools and to reign in the massive resistance to prayer and Bible reading injunctions, conservatives created alternative schools. Sometimes the mix of religious and racial motives were obvious, as in the wholesale movement by whites into "private" segregation academies in such areas as Prince Edward County, Virginia, and the Mississippi Delta region, often financed by government through voucher programs or more clandestine means. In the words of Mississippi Citizen' Council staffer Medford Evans, such schools would preserve an "island of segregation" just as "monasteries saved the Greek and Roman classics" during the "Dark Ages." By 1968 forty-two segregation academies were receiving tuition vouchers from the state of Mississippi. By 1973, after more rigorous court enforcement of desegregation, there were 125 such schools, many with enrollments in the thousands.[43]

Other private schools founded during this time were less obviously race-based and more clearly driven by religious concerns. Evolution, sex education, the somewhat vague but powerful notion of "secular humanism," and other factors drove many families away from public education. Many conservatives gave up, at least for the time being, on the idea of transforming the public school and sought instead "to restore power to local evangelical communities by creating a parallel educational culture." The schools they founded were not typically sponsored by denominational bodies but by local churches or even just a few individuals. Many of them joined a Christian school association that could provide accreditation, professional training, insurance packages, legal assistance in the event of conflict with the state, and entry into a network of like-minded schools. The largest of these organizations, the Association of Christian Schools International (ACSI), was formed in 1978 as a merger of three regional organizations. In 1967, these three parent organizations had a combined membership of 102 schools enrolling 14, 659 students. By 1973 they had 308 schools. Over the next three decades enrollment skyrocketed:

Table 4.1   Growth of ASCI

Year	Number of ACSI Schools	Total Enrollment
1983	1,900	270,000
1989	2,347	340,626
1993	2,801	463,868
2000	3,849	707,928
2005	3,957	746,681

The second largest organization, the more aggressively separatist American Association of Christian Schools (AACS), was founded in 1972 with 80 schools enrolling 16,000 students. It experienced similar growth in the 1970s but by the mid-1980s had stagnated and in more recent years even declined a bit:

Table 4.2 Growth of AACS

Year	Number of AACS Schools	Total Enrollment
1983	1,100	160,000
1991	1,200	187,000
2004	1,050	175,000

A third organization, Christian Schools International (CSI), was historically a Dutch Reformed group, but it expanded in the 1980s to include many of the new pan-denominational Christian schools. By 1987 it had 400 member schools and 1,200 nonmember schools that used some of CSI's services but could not officially join due to the reformed theological stance CSI required.[44]

Such rapid growth in the private sector is astounding, especially in light of the overall decline in number of children in the nation during this period and the concomitant decline in enrollment nationwide in all other types of schooling. But, in fact, the growth of private Christian day schools was even more profound than the figures of the leading umbrella organizations indicate. Many Christian schools founded during the 1970s and 1980s were intentionally unaffiliated with any group and proudly unaccredited by the state. As pastor Rodell Bledsoe of the Bible Baptist School in Bismark, ND, explained,

> We don't want approval, because we feel it's a matter of state control. Jesus said in Matthew, Chapter 16, "I will build my church, and the gates of hell will not prevail against it." We believe the head of the Church is Jesus Christ, and if I let the Sate become the head of the church, then I will be removing the Lord from His position, and this Church is definitely built on the Lord, Jesus Christ.

Given this antipathy toward registration or accreditation, no one knows how many independent schools were actually formed. One scholar estimated that by 1984 approximately 6,000 independent schools were in existence, while another thought the figure was closer to 15,000. Movement advocates placed the number even higher, at perhaps 20,000 to 25,000. Whatever the precise figure, it is abundantly clear that a large-scale shift in how conservative Protestants thought about and practiced education took place in these decades. While most remained in public schools, a growing and committed minority agreed with Christian educators like Kenneth Gangel that "the children of God deserve something better than pagan public education. When we give our sons and daughters to the secular system we invite the values, standards, and errors of a godless culture to penetrate their spirits." Indeed, in some congregations parents who kept their children in public schools were seen as operating outside of God's will. By the 1990s the whole issue had become quite divisive in some Evangelical circles, with some arguing that the schools were "doing more harm now to the country than any single thing except perhaps the popular media" and that Christians must get out immediately, while others claimed that Christians needed to stay in the system to serve as "salt and light" to their non-Christian neighbors.[45]

But for some conservative Christians, private schools were not the answer. While agreeing with the critique of public education that had led thousands of conservatives into Christian day schools, some parents could not accept this solution. Reasons for dissatisfaction with private schooling varied: some families couldn't afford the tuition; some disagreed with the theology their local school(s) espoused; some had personality conflicts with principals or teachers; some, especially those with special needs children, felt that the private school couldn't adequately address their child's individual circumstances; some believed that the Bible gave responsibility for education to parents only; and some, especially mothers, simply wanted to spend more time with their children. For any combination of these reasons and, doubtless, others besides, some conservative Christian parents began to give homeschooling a try. The circumstances were right. By the late 1970s many conservatives lived in comfortable suburban homes that could easily accommodate a homeschool. Many housewives were well educated and committed both to their children and to staying at home. A Gallup poll in the mid 1970s showed that 60 percent of housewives did not want to work outside the home. Most listed "my children" as their biggest source of pride. These were empowered women, politically astute and activist. Housewives, as we have seen, "formed the backbone" of most pro-family movements. If such women as these could protest, organize voters, conduct study groups, and lead Bible studies and women's clubs at their churches, could they not teach their own children how to read, write, and cipher? Many decided they could.[46]

Some critics have suggested another reason some conservative Protestants may have turned to homeschooling over private schools in the early 1980s. Beginning in 1970 the Internal Revenue Service, responding to a Mississippi court's decision in *Green v. Connally*, changed its longstanding policy of granting tax-exempt status to private schools that discriminated against minorities. The Supreme Court upheld this shift of policy in *Coit v. Green* (1971), *Runyon v. McCrary* (1976), *Prince Edward School Foundation v. United States* (1981), and, most famously, *Bob Jones University v. United States* (1983). Many conservative Christians who had formerly avoided politics for the most part became enraged, especially when the IRS issued more stringent guidelines in 1978 that would have looked not only at formal admissions policies but at actual enrollment. Paul Weyrich went so far as to claim that "what galvanized the Christian community was not abortion, school prayer, or the ERA.... What changed their mind was Jimmy Carter's intervention against the Christian schools, trying to deny them tax-exempt status on the basis of so-called de facto segregation." For the first time a pan-denominational coalition of Christian conservatives united in vocal opposition to the Federal Government's attempt to regulate them. Tens of thousands, many of whom had never done so before, wrote letters to congress protesting the IRS initiative. The response was so overwhelming that Congress gutted the new measures, easily passing an amendment authored by Jesse Helms forbidding the use of any federal funds to challenge the tax-exempt status of any private school.[47]

Nevertheless, a handful of schools were prosecuted, and two cases at least made it to the Supreme Court, the most famous of which is *Bob Jones v. U. S.* (1983), where the Supreme Court in dramatic fashion upheld the IRS' lifting of the tax-exempt status of Bob Jones University and the Goldsboro Christian School in North Carolina. Given the shrill rhetoric and extensive coverage of this issue in the press, it is understandable that some commentators have seen herein a broader shift with implications for homeschooling. Linda Dobson, for example, has argued:

> In the 1980s, changes in the tax regulations for Christian schools forced the smaller among them to close down by the hundreds. Suddenly, the parents of the students attending these schools were faced with a choice between government school attendance and homeschooling. For many, this really wasn't a choice at all, and these Christian families became part of a large second wave of homeschooling, joining earlier homeschoolers and boosting the numbers to record

highs. Christian curriculum providers, already well-established businesses that had just lost a large chunk of their original market, followed the money and easily courted the new market of homeschooling parents.

It's a plausible claim, with deliciously scandalous implications: segregation academies were hunted down by the IRS and forced to close, leading to a mass movement of racist Christians into homeschooling where they brought their segregationist curriculum (mostly Bob Jones and Abeka, the two most popular curriculum packages among conservative Christian homeschoolers) with them. It explains the striking correlation between the *Bob Jones* court decision, the slowdown in AACS school growth in the mid-1980s, and the explosion of homeschooling at precisely the same time. But it is simply not true. Of the hundreds of fundamentalist schools that closed in the mid-1980s and 1990s, some of whom no doubt were segregation academies, only a small handful did so due to pressure from the IRS. By the 1980s the IRS had no money and the Reagan administration no will to go after discriminatory schools. Most of the schools that closed did so because they were founded "with more enthusiasm than resources and leadership." By 1983 many of these schools were simply imploding, stranding the few families who patronized them. Homeschooling was a lifesaver for these families. Furthermore, many school closings were the consequence, not the cause, of the shift to homeschooling among their clientele. By the mid-1980s homeschooling was becoming increasingly popular among religious conservatives, and thousands of them pulled their children out of Christian day schools to enroll them in homeschools.[48]

Many families who experimented with homeschooling in the early 1980s ran into all sorts of trouble. Their own churches often frowned on the practice (especially if the church ran its own school). Their extended family members often thought they were crazy. Media outlets and especially public school people feared for the safety and future of homeschooled children. In many states, extant laws made the practice either illegal or of dubious status. Conflict between families and government was sometimes ugly. In the coming chapters we will examine in more detail how homeschooling emerged from the surreptitious underground to win mainstream acceptance. But for now let me draw many of the themes I've been discussing together to summarize why homeschooling happened.

Firstly, homeschooling happened because the countercultural sensibility became the American sensibility. As historian Bruce

Schulman put it, "During the Seventies, the forces of God and the forces of Mammon refused to show deference to established leaders and institutions." Having rejected the mainstream, denizens of both left and right looked for personal fulfillment within a small, alternative community. Social and political changes of the second half of the twentieth century made bedfellows both of radical leftists who wanted nothing to do with conventional America and conventional Americans who wanted nothing to do with a country that in their view had sold out to the radical left. This anti-institutionalism led some on the right to reject not only public schools but Christian schools as well. Some even rejected organized Churches altogether, claiming a New Testament mandate for the establishment of informal "house churches." As one teen baldly put it, "I love God...but hate the church."[49]

Secondly, homeschooling happened because of suburbanization. Its deracinated and media-saturated environs incubated the alienation that led so many young people to challenge the system by leaving it, founding communes, and pioneering homeschooling. It facilitated the segregation of the population by race, income level, age, number of children, and cultural style, thus feeding the American hunger for privacy. Though built and sustained largely by government, it was a breeding ground for libertarian sentiment and antigovernment activism. It gave homemaking women a set of causes to fight for and a base from which to operate their campaigns. And not least, it provided some of these women with the physical space they would need to teach their kids.

Thirdly, homeschooling happened because of the American cult of the child. The progressive left had long harbored romantic ideals of child nature, born of Rousseau and come of age in the progressive education movement of the early twentieth century. Countercultural leftists inherited this outlook, and when they had children their instinct was to liberate the kids from what they took to be the deadening effects of institutionalization by keeping them at home. And the countercultural right, despite ostensibly conservative and biblical theological commitments, had basically the same view. In earlier chapters we noted the difficulty American Protestantism has had in preaching and maintaining the doctrine of original sin. Successive waves of revival have bequeathed a deeply ingrained belief in the freedom of each person to choose whether or not to follow Christ instead of the fatalistic notion of the will in bondage to sin. If asked, many conservative Christians will say they believe in original sin, but at the deepest level they tend to think of their children as precious gifts of

God, full of potential, not as vipers. Just as conservatives have adapted to the culture's commercialism, its backbeat rhythms and glossy self-help style, so they have embraced the romantic cult of the child. Sociologist Mitchell Stevens concluded after years of careful study of conservative homeschoolers that their core belief about children is that "deep inside each of us is an essential, inviolable self, a little person distinctive from all others." The words of one of the best-loved children's songs of Christian music power couple Bill and Gloria Gaither perfectly captures this very mainstream idea,

> I am a promise, I am a possibility,
> I am a promise with a capital "P"
> I am a great big bundle of potentiality.
> And I am learnin' to hear God's voice
> And I am tryin' to make the right choices.
> I'm a promise to be
> Anything God wants me to be.

During the late 1970s and early 1980s, the "decade of nightmares" about prowling sexual predators, child-molesting teachers, debauched youth culture, and occult brainwashing, many parents sought shelter in the safety of the home to nourish the promise of their children.[50]

Fourthly, homeschooling happened because of changes both in public schooling and in families during the second half of the twentieth century. As public schools grew larger, more bureaucratic and impersonal, less responsive to parents, and less adaptable to individual or local cultural variations, many families felt increasingly alienated. The loss of a discernibly Christian school culture, courses on subjects some parents found offensive, curricula that undermined home values, exposure to children of other races, religions, and family structures, and many other factors only added to this alienation. Many parents began to connect the dots between changes in school culture and broader changes in American families, such as the dramatic increase in divorce, out-of-wedlock births, abortion, single-parent homes, and calls for gay rights. Fearing for their own children's futures, they pulled out. For some of them, ostensible Christian schools presented similar problems—carousing students, inflexible administration, objectionable curriculum, not to mention the cost.[51]

But homeschooling was not simply the inevitable result of these broad social forces. It happened because real people engaged in a multipronged effort to challenge the dominant approach to childhood education. Intellectuals articulated the vision. Parents tried it out.

Lawyers and politicians worked to smooth the way. Organizations emerged to facilitate networking among homeschoolers and eventually to sort them into competing tribes. Entrepreneurs and eventually corporate conglomerates rushed to meet the demand of the growing movement for curriculum materials. In the next chapters we will deal with all of these factors as we examine how homeschooling progressed in just a few short years from being the furtive activity of a few radicals to a mainstream option chosen by millions of Americans, endorsed by government, praised by media, and accepted as legitimate education by the most elite colleges and universities.

Chapter Five

Three Homeschooling Pioneers

John Singer had always had a problem with authority. As a boy growing up in Germany, he had been compelled to participate in the Hitler Youth program but was expelled from an SS school due to his incorrigibility. In 1945 his mother, a Mormon, divorced her husband and moved with her son to Utah. In 1964 Singer married Vickie Lemon. Very quickly they started having children, seven in all, and raised them on a farm in Marion, Utah. Singer had always been an outsider, but by the 1970s his religious and political views had put him so at odds with the Mormon hierarchy that they excommunicated him. In 1973 he removed his children from the Utah public schools after his daughter had come home one day with a textbook that celebrated Martin Luther King, Jr. as a patriot and featured a photograph of blacks and whites together. Singer was arrested for this action but permitted by a Utah court to homeschool his kids so long as they were tested twice a year and received an annual psychological evaluation at the Singer home. After the first of these, Singer told the psychologist never to come back. For five years school officials and law enforcement left the family alone.

In 1978 Singer married Shirley Black, who was already married with four children. Black and her children came to live on the farm with Singer, his first wife, and their brood. In October of that year, Black's first husband was awarded custody of her four children at a divorce hearing, but Singer would not give them up. On October 19 authorities surrounded Singer's home, calling for the release of the children, but the family stayed barricaded inside, where they had stockpiled plenty of food, firearms, and ammunition. The siege lasted for three months. On January 18, 1979, Singer left the house to do a few chores. After getting his mail he was suddenly ambushed by ten police officers on snowmobiles. They told him to surrender his weapon, but instead he drew it and ran. Singer was shot at least six times in the back and killed.

Singer's two wives were incarcerated and the children placed in shelter homes. A civil suit was filed on behalf of the Singers (*Singer v.*

Wadman, 1982), but a federal judge prevented the trial from going before a jury. The freakish nature of the story, and especially the lack of resolution, shocked Utah and made Singer something of a hero to antigovernment fringe groups. Singer's first wife Vickie eventually regained custody of her children and continued to educate them at home. Nine years after Singer's death, she had a vision that her husband would be resurrected and that this event would presage the second coming of Christ. The Singer clan together watched a videotape of John's funeral, which whipped them into a religious frenzy. Then things got ugly. Addam Swapp, who had married both of Singer's daughters, dynamited a nearby Mormon church's skate center. The family then retreated yet again into their compound. After a thirteen-day siege and the death of a Utah Department of Corrections officer, police stormed the compound. Swapp and Singer's son John Timothy (who fired the shot that killed the officer) were arrested and sent to prison. In 1992 a made-for-television movie called *Children of Fury* dramatized the events.[1]

On the other side of the country, in Amherst, MA, only months before Singer would be gunned down, the son of Peter and Susan Perchemlides was charged with truancy. It had been a long time coming. Peter, a biochemist with a Ph.D. from Duke University, had left his job in Boston to move the family to Amherst so Susan could pursue graduate study in social and political theory at the university there. Back in Boston the couple had homeschooled their three boys for four years. When they moved to Amherst they enrolled them in public school, where the two older boys performed well. But their youngest son Richard, a second grader who had been "a free, confident child" became "shy, unsure, self-conscious, and discouraged over academic achievement." To try to help their son, Peter became very involved in the local parents' council. He quit midyear, however, because "there was no opportunity for parental input, only for rubber-stamping decisions." Peter became convinced that the school was intractably committed to a hidden curriculum of "conformity, anti-intellectualism, passivity, alienation, classism, and hierarchy." Susan likewise complained that the school had a fatal tendency to "break down and categorize curriculum, and they break down and categorize children too."

So when the time came to register Richard for third grade in the fall of 1977, the Perchemlideses decided to teach him at home. They met with Superintendent Donald Frizzle, who gave them a form titled "Private School Application for School Committee Approval." The family submitted the form in September, proposing the Spalding program (a phonics-based approach) for language arts and other curricula covering math, music, art, and ecological subjects. In October their proposal was rejected. Four reasons were cited. First, it was alleged that the parents did not have the appropriate training or background to educate their children. Second, the curriculum was not properly sequenced or keyed to the child's development and capacity. Third, no group experiences were provided and the school committee held these "essential to a child's personal and intellectual growth." Finally, the committee determined that Richard's poor performance in second grade was the result of his prior home instruction in Boston, not his school experience.

The Perchemlideses appealed and submitted a much more detailed plan. This second effort was never read and was summarily rejected. Though no rationale was provided to the parents, committee meeting minutes reveal that one member thought approval would set a bad precedent. A second member distrusted the motives of anyone who would do something as un-American as take a child out of public school. Another thought that home education should only be a last resort in situations of extreme physical handicap. Yet another thought it would be impossible to monitor the child's progress.

Despite the rejection, the Perchemlideses did not send their son to school. After months of threats, Superintendent Frizzle finally brought truancy charges in April of 1978 and asked for warrants to arrest the parents. To avoid arrest and win the right to teach their child at home, the Perchemlideses sued the school district. They contacted the Western Massachusetts Legal Services and were fortunate to find a young Northampton lawyer named Wendy Sibbison who was willing to take on the case for "a pittance" because she found the issue intellectually stimulating. The Cambridge Center for Law and Education also worked *pro bono* for the Perchemlides family.

At trial, Sibbison argued that the Perchemlideses had a constitutional "right to privacy" that included the right to educate their children as they saw fit, a right that could only be abrogated if there were a "compelling state interest" to do so. District judge Greaney accepted this argument. In November of 1978 he decided against the school district. His opinion stated, "Under our system, the parents must be allowed to decide whether public school education, including its socialization aspects, is desirable or undesirable for their children." The school committee, he argued, had no right to take into consideration parental motivation, socialization of the child, fears of setting a bad precedent, or any other factors that do not bear on the actual content of the curriculum plan submitted by the Perchemlideses. And as for the content of the plan itself, the judge agreed with the many expert witnesses called by the prosecution who, having analyzed the curriculum plan submitted by Susan, found it "the equivalent of a first rate private academy." Judge Greaney opined that homeschooling was legal in Massachusetts provided three conditions were met: the parents were competent (the judge did not spell out qualifications), the subjects required by law were taught for the number of hours and days the law mandated, and there was some sort of accountability measure to ensure adequate progress.

The family immediately resubmitted their plan and Superintendent Frizzle approved it. "I am not opposed to home education," he said, "I was merely trying to safeguard the child's interest. The judge's decision on what to consider and what to disregard sorted it out for me." Altogether the school district lost $7,000 in legal fees and quite a bit more in public esteem, for the case had garnered an enormous amount of media attention. In every instance the case was given a David and Goliath spin—a noble and persecuted family fighting back against an obdurate and cruel bureaucracy. Legal scholar Stephen Arons, for example, dissected the case in law journals and popular outlets, issuing stinging barbs against school districts that "see home teaching as a threat to public schooling" even as they ignore "thousands of dropouts." The case led homeschool activist John Holt to advise school administrators not to take on homeschoolers, because "the media invariably side with the family, and the schools end up looking like bullies. If you win or lose, you still lose."[2]

What is perhaps most significant about both the Singer and Perchemlides stories is that by the late 1970s they were not anomalies. Earlier decades had seen isolated examples of conflicts between homeschooling families and local school officials, but as we have seen, home instruction chosen as a deliberate alternative to institutional schooling had become so rare by the mid-twentieth century that occasions for conflict were few. Every so often a new agrarian like Ralph Borsodi might shake the dust of the city off his sandals, move his family to the farm, teach them at home, and write a book about it (*Flight From the City*, 1933). Occasionally a mother like Rita Scherman (*A Mother's Letters to a Schoolmaster*, 1923) or a father like William Barrett (*The Home Education of a Boy*, 1950) might get fed up with standardization and bureaucracy and pull out a child. Sporadically a family like the Cardens of Lavergne, TN, might wish for something more rigorous than what was provided by their rural district and order the Calvert curriculum by mail. Though there were isolated cases of conflict with school authorities, most of the times such folk were left alone.[3]

By the late 1970s, however, more and more families were deciding against institutional schooling, increasing the possibility both of conflict with local schools and of reaching a tipping point that might transform the decisions of a few into a full-fledged movement. By the 1970s such books as Hal Bennett's *No More Public School* (1972) and Howard Rowland's *No More School* (1975) were increasingly common. Futurist Alvin Toffler (*Futureshock*, 1970) presciently noted homeschooling as an up-and-coming trend. Veteran Montessori teacher Rosa Covington Packard encouraged families to apply Montessori's methods in the home (*The Hidden Hinge*, 1972). In the previous chapter we saw why more and more families on both the right and left were considering anew the notion of the home as school. In the next three chapters, by examining three overlapping domains, we look more closely at how the movement was born and grew. In this chapter we consider three intellectuals, writers, and organizers whose advocacy and activism catalyzed various wings of the movement. In chapter six, we look at the organizations and curriculum providers that facilitated the movement's growth and differentiation. Finally, in chapter seven, we examine the legislative and legal history of the homeschooling question.[4]

Most of the advocates for homeschooling only turned to it after they had become convinced that institutional schooling, especially public education, was bad for children. Susan Perchemlides' conviction that school was killing her son's creativity and homogenizing his soul echoed sentiments that had, by the late seventies, become commonplaces of best-selling school critiques written by left-leaning intellectuals and activists. Such writers as Jonathan Kozol, George Dennison, and Peter Marin produced a flood of books and articles condemning the numbing public school bureaucracy and urging the nation to set children free. At the same time, John Singer's fear that his children would imbibe attitudes and ideas that challenged his religious sensibilities was shared by many on the right who felt that the nation's schools had traded in religion for secular humanism and academics for values clarification. Timothy LaHaye's Family Life Seminars, founded in 1971, taught families "how to resist the corrosive influences of liberal culture," primarily by leaving public education and establishing separate Christian schools, ideally staffed by teachers trained at his Christian Heritage College (founded in 1970). LaHaye and many other conservative Christian leaders produced, and continue to produce, a torrent of literature indicting public education for training children "to be anti-God, antimoral, antifamily, anti–free enterprise and anti-American," as LaHaye's best-selling 1983 book *The Battle for the Public Schools* put it.[5]

Jeremiads against public education weren't just coming from the radical left and right. Criticisms of all sorts were heaped upon the public schools in the postwar decades from more mainstream sources as well. The best-selling *Myth of the Hyperactive Child* (1975) decried the increased use of psychological testing, sedatives, and clandestine record-keeping. Famed psychologists Kenneth Clark and Edgar Friedenberg warned of the "vestibule adolescence" schools produced by sequestering children away from adult experiences and struggles. Libertarian Milton Friedman, liberals John Coons and Stephen Sugarman, and socialist Christopher Jencks all advocated family control of education through voucher plans. Large numbers of people from all walks of life and ideological orientations left public education and joined free schools, religious academies, progressive cooperatives, and many other private options. It is not surprising that in all this frenzied critique of public education and rush to find alternatives, some critics hit on homeschooling as the answer. In what follows, we look at three writers and activists who exerted profound influence on the homeschooling movement. Other "pioneers" of course served important roles in advocating and organizing homeschooling, but the subsequent development of the movement in all of its contradictions and conflicts is best understood in light of the early contributions of John Holt, Raymond and Dorothy Moore, and Rousas J. Rushdoony.[6]

John Holt

No one who knew him would have predicted Holt's prominent role as homeschooling champion, theorist, and organizer. Though born to New England wealth, his relationship with his parents was near nonexistent. Holt's sister recalled how both she and John "felt we were an unwanted burden to our parents." Holt spent his childhood studying with private tutors and at some of the most prestigious boarding schools in the country. Then it was off to Phillips Exeter for high school and Yale for college, where Holt graduated in 1943 with a degree in Industrial Engineering ("whatever that means," he would later remark). Holt was clearly embarrassed by the pretentiousness of this background; he never would name the schools he had attended. He was also quite clear that he found most of what he experienced in these places worthless despite his stellar academic record.[7]

After college, Holt served as a lieutenant aboard a submarine in the Pacific for three years, an experience he recalled fondly as "the best learning community I have ever seen or been a part of." But the atomic bomb profoundly disturbed him, so much so that he joined a group seeking peace through one world government called the United World Federalists. For six years Holt played the political troubadour, canvassing the country and preaching the gospel of global peace. By 1952 he had tired of this work and took a year off to bike around Europe. Back in the United States, Holt's sister, noting his interest in and facility with her children, encouraged him to visit the Colorado Rocky Mountain school, a coeducational free school emphasizing manual training. Holt taught there for four years and then moved back to the east coast, where he got a job teaching at a selective private school in Cambridge, MA. Here he met Bill Hull, an intellectual comrade. For the next seven years Holt and Hull observed one another's classes, taking meticulous notes. These notes became the basis for the books that made Holt's name.[8]

Holt's first book, *How Children Fail*, was rejected by several publishers before finally seeing the light in 1964. It quickly became a bestseller and remains popular as a classic statement of 1960's era school critique. Holt's basic contention, richly illustrated by anecdotes from his classrooms, was that compulsory schooling destroys children's native curiosity and replaces it with a self-conscious and fearful desire to please the teacher. Kids learn not rich subject matter but skills necessary to pass tests and charm authorities.

In 1967 Holt produced a sequel volume, *How Children Learn*. Again drawing on his extensive notes from classroom observations, Holt continued his attack on formal education. He also described in very positive tones the natural education children receive in the home before schooling takes over. Together, *How Children Fail* and *How Children Learn* have sold over 1.5 million copies. Holt was able to succeed where many similarly minded reformers failed largely because of his rhetoric. The contrast between natural, safe, and caring home education and compulsory, fear-driven, test-crazy school was presented in clear language, with compelling anecdotes and great passion. Holt never came across as bombastic or curmudgeonly person even at his most critical moments; his language was always simple and free of academic jargon. Holt's steady supply of anecdotes—some humorous, some tragic, all engaging, make his books difficult to put down. Parents were especially moved by them, and Holt received and responded to thousands of letters filled with stories of children abused by the educational system. By the 1970s, reformer Peter Marin could say without exaggeration that Holt was "the country's most popular education writer."[9]

He was not, however, the country's most popular teacher. Holt was fired from several schools for his refusal to accommodate administrative needs. He tried to run his fifth grade math class without recourse to any sort of assessment or evaluation and often suggested pedagogical reforms that scandalized even progressive private school administrators. But by 1968 his books had made him so well known that better offers started coming his way. That year he served as a visiting lecturer at the Harvard Graduate School of Education. In 1969 he did the same at the University of California at Berkeley. At this point he still believed that schools could be transformed into positive resource centers for all. His optimism about the possibility that schools could be made into free spaces changed in 1970 upon a visit to Austrian-born social critic Ivan Illich's Centro Intercultural de Documentación (CIDOC) at Cuernavaca, Mexico. Illich was a renegade Catholic priest trying to use CIDOC to prepare Catholic missionaries not to modernize third-world peoples but to embrace their subsistence economy and simple technologies. Like Holt, he was critical of modernist education, but he provided for Holt a broader intellectual context from which to shape a critique not only of schools but of all technocratic institutions that imprisoned the human spirit in bureaucratic cages.[10]

In the mid-1960s Holt had been seen as a mainstream figure, frequently contributing articles to such periodicals as *Life*, the *Saturday Evening Post*, *Redbook*, even the *PTA Magazine*. He traveled widely, lent his voice and enthusiasm to the free school movement then in full bloom, and charmed many middle-class parents with his stories in lectures across the country. Then, as the sixties became the seventies, Holt radicalized like so many others. He became an outspoken critic of the Vietnam War, refused to pay taxes, and supported draft dodgers and student free-speech rights. He even turned down an honorary doctorate from Wesleyan University in 1970, arguing in his commencement rejection speech that colleges were "among the chief enslaving institutions" in America. Holt's 1972 *Freedom and Beyond* showcased both his growing awareness of the larger context within which children grow up and his increasing disaffection with schooling as such. He began to doubt that schools, *any* schools, could challenge the racism, classism, and economic reductionism of modern life. Echoing his friend Illich, who in 1971 had published the influential book *Deschooling Society*, Holt argued that children needed to be liberated from schools altogether. His 1974 *Escape from Childhood* extended this line of thought to argue that children should be granted eleven basic rights, including the right to earn money, to sue and be sued, to choose their own guardians, to vote, travel, and learn in whatever way they wished. By this time he had alienated most of his original fans, many of whom were disappointed with his shift from what one reviewer called his "incisive" use of "close-up first-hand experiences" to this new "preachy and unthoughtful puff on children's rights."[11]

By the mid-1970s Holt had gone from being an educational celebrity with mammoth sales and a full docket of public engagements to a fringe figure even among radical critics of schooling, most of whom were outraged at his categorical dismissal of all schools, even free schools. But then in 1976, almost by accident, Holt found himself at the forefront of a movement that would dominate his time and energy for the remainder of his life. That year he published *Instead of Education: Ways to Help People Do Things Better*. The book is a largely forgettable hodgepodge of possible alternatives to institutional schooling: voluntary learning centers, reciprocal learning environments where people of all ages come to learn from one another, informal educational networks between private individuals, and so forth. But one of his ideas, suggested almost whimsically, changed his life. Late in the book Holt mused on the possibility of creating a "new Underground Railroad" to help children escape from schools. Children were enslaved in schools, and courageous individuals ought to engage in any means possible, legal or not, to liberate them.

The suggestion was noticed by some of the communards, homesteaders, and other political outsiders who had been turning to homeschooling in the 1970s. Sensing in Holt a possible ally, several of them contacted Holt to tell him about their experiences of homeschooling their children and especially about their struggles with local school authorities. Holt found this underground subculture fascinating and decided to do whatever he could to support their actions. He became an opinion broker and networking agent, creating the first newsletter in the country aimed at homeschoolers. The first issue of *Growing Without Schooling (GWS)* was published in August of 1977, a four page, typewritten collection of Holt's ruminations, homeschooling resources, school horror stories, and letters Holt had received from some homeschoolers. Though subscriptions were few at first, the beginnings of a homeschool network were born as previously isolated individuals, many of whom thought they were the only people doing what they were doing, became aware of one another, exchanged stories and tips, developed legal strategies, and started thinking of themselves as the vanguard of a true grassroots movement. Very quickly social networks were formed, operating under the moniker HOUSE, which at first stood for "Home Oriented Unschooling Experience" but was later changed to "Home Oriented Unique Schooling

Experience" out of respect for those in the movement who were uncomfortable with the anarchical flavor of the word "unschooling." Holt here was advancing his "nickel and dime theory about social change" wherein a small group of early adopters were clearing a path that more and more people would gradually travel.[12]

Holt's fame, rhetorical skill, and tireless activism quickly made him the de facto leader of the homeschooling movement. His leadership was one of sacrifice and service. Whereas many later homeschool leaders would profit greatly from their efforts, Holt spent tens of thousands of dollars of his own royalty monies and lecture fees to bankroll court cases, sustain *Growing Without Schooling*, and travel around the country speaking, demonstrating, and witnessing in court on behalf of homeschoolers. Slowly the mainstream media began to take notice of this movement, especially as court cases like those discussed at the beginning of this chapter became increasingly common.[13]

The tipping point came in December 1978. The Perchemlideses had just won their court case in November. A second and similar case in Iowa involving Bob and Linda Sessions and their right to educate their child at home had also just been won. In December, *Time Magazine* ran an article on the entire movement, the first of its kind in a major American weekly. A few days later John Holt appeared on *The Phil Donahue Show* with the Sessions family for a rousing Donahue-style discussion of homeschooling. The show had an immediate and dramatic impact on the scope and prestige of homeschooling. Even though Donahue's audience was largely critical of what Holt and the Sessions family were doing, it was clear that Donahue himself was intrigued, and Holt's sincerity, humor, and penchant for folksy examples and homespun wisdom won many over. This was not the John Holt of radical kiddie lib or anarchist revolution, but the lovable Holt of the late 1960s telling heartwarming stories of family life and heartbreaking stories of institutional sins against childhood.[14]

After the broadcast, subscriptions to *GWS* tripled to 1700. Other media outlets picked up the story. Speaking offers came pouring in for Holt, as did some 10,000 letters. He later wrote that out of all the correspondence, "only eight letters were critical and/or hostile," noting particularly that hundreds of supportive letters were written by teachers. Many of the first wave of homeschooling families trace their inspiration back to that first Donahue show. Linda Dobson, who would eventually become news editor for *Home Education Magazine* and pen the popular how-to book, *The Art of Education*, recalls how at the very time when she was growing increasingly frustrated with her eldest son's experience in kindergarten in New Jersey "Phil Donahue was smart enough to have John Holt on his show.... The rest, as they say, is history." Joyce Kinmont, a pioneer homeschooler in Utah and key organizer for Mormon homeschooling, recalls fondly how Holt's appearance validated what she was trying to do in the eyes of many of her disapproving and suspicious coreligionists. Kinmont and her family of seven children were invited by Holt to join him on his second appearance on Donahue's show. "The station flew our entire family to Chicago, plus a nanny for the baby.... That was our most exciting field trip!"[15]

One of Holt's most appealing qualities was his willingness to listen to and make common cause with people from a wide range of ideological perspectives. Very quickly he became aware that the movement he had discovered was peopled not only by leftists who resonated with his concerns for child liberation, but also by very many conservative Christians who maintained traditional notions of patriarchal authority, childhood obedience, and biblical morality. Holt developed collegial relations with Christian homeschool leaders, recommending their resources in his magazine and making common cause with them in legal proceedings and political organizing. Holt acknowledged, as he put it in the second issue of *GWS*, that people who reject institutional schooling do so for various reasons:

> Some may feel that the schools are too strict; others that they are not strict enough. Some may feel that the schools spend too much time on what they call the Basics; others that they don't spend enough. Some may feel that the schools teach a dog-eat-dog competitiveness; others that they teach a mealy-mouth Socialism. Some may feel that the schools teach too much religion; others that they don't teach enough, but teach instead a shallow atheistic humanism.... What is important is not that all readers of *GWS* should agree on these questions, but that we should respect our differences as we work for what we agree on, our right and the right of all people to take their children out of schools.... We will try to be as useful as possible to *all* our readers, whether or not we agree with them on all details.[16]

One group that Holt was especially attuned to was the Seventh-Day Adventists. In the first issue of *GWS* he noted the existence of the "well established, respectable, and very extensive correspondence school" called the Home Study Institute, which "seems to be run by, or somehow connected to, the Seventh Day Adventists." He noted with interest that "the Adventists have had a good deal of experience in bucking compulsory attendance laws... in short, these folks may already know a great deal that we need to find out." One Adventist couple who did indeed know more than Holt did and with whom Holt quickly established a working relationship was Raymond and Dorothy Moore.[17]

Raymond and Dorothy Moore

Seventh-Day Adventists, from their founder Ellen G. White onward, have always articulated a theology of the family that would mesh well with homeschooling. White wrote voluminously on the topic of the family in Adventist periodicals, pamphlets, and published books throughout her seven decades of church leadership from 1844 to her death in 1915. In hindsight, portions of her work read almost as a rallying cry for homeschooling, anticipating by several decades the rhetoric that would eventually enjoy wide popularity among all sorts of Christians. White habitually asserts that God, in His wisdom,

> has decreed that the family shall be the greatest of all educational agencies. It is in the home that the education of the child is to begin. Here is his first school. Here, with his parents as instructors, he is to learn the lessons that are to guide him throughout life.

White especially emphasizes the role of the mother, who "must ever stand pre-eminent in this work of training the children." She chides those Adventist parents who "cast off their God-given responsibility to their children, and are willing that strangers should bear it for them," believing that such an attitude opens the door for "Satan and his host" who "are making most powerful efforts to sway the minds of the children.... The world today is destitute of true goodness because parents have failed to gather their children to themselves in the home."[18]

While no records exist documenting how many Adventist families chose to homeschool their children, it is clear that Adventists have long considered this a viable option, especially in a child's early years. In fact, one of the earliest court cases involving the rights of parents to teach children at home involved an Adventist family. Marjorie Levisen's truancy conviction was reversed by the Illinois Supreme court in 1950, which held that her daughter's at-home instruction did qualify as private schooling under Illinois law. In the majority decision, Justice Crampton noted that the Levisens

> are Seventh Day Adventists in religion, believing that the child should not be educated in competition with other children because it produces

a pugnacious character, that the necessary atmosphere of faith in the Bible cannot be obtained in the public school, and that for the first eight or ten years of a child's life the field or garden is the best schoolroom, the mother the best teacher, and nature the best lesson book.

Justice Crampton's description is an apt summary of the hallmarks of Adventist educational philosophy. And it describes perfectly the approach to education of Raymond and Dorothy Moore.[19]

Raymond Moore was born in 1915 to pious Baptists in California. Dorothy was born the same year to Methodists in North Dakota. When Raymond was three years old his mother died and his father became an Adventist. Raymond himself was baptized into the Adventist church at age twelve. Dorothy did not convert until college. The couple met at the Adventist Pacific Union College, where Raymond earned a degree in English education and communication and Dorothy took a degree in education. They married in 1938 and both worked in the California public schools. Raymond joined the army during World War II, where he rose to the rank of major, serving on General MacArthur's staff in Manila. When the war ended, he returned to school, earning a Ph.D. in developmental psychology and teacher education in 1947 from the University of Southern California. Raymond was immediately lapped up by the Adventists and held a series of administrative posts within the denomination's school system until 1964. During this time Dorothy reared nine children—two biological and seven "adopted," and earned an M.A. from Andrews University in 1959, writing a thesis on remedial reading programs in Adventist schools.[20]

Dorothy had long been deeply influenced by Ellen G. White's ideas about child rearing. In 1937, while teaching remedial reading in Southern California schools, she noted that many of her students had been placed in school very early. Convinced from her reading of White that mother and home are the best early educators, she began to formulate the conviction that early formal training damaged kids. In the late 1940s, as Dorothy began to teach her own young children at home, she grew increasingly incredulous toward the push to get kids in school so early. Raymond later recalled that it was due to her concerns that "I started researching this." The results of his researches were probably the most significant catalyst in homeschooling's emergence as a national movement.[21]

The Moores had founded a fledgling research institute in 1961 called the Moore Foundation, but it didn't really get off the ground. Raymond was working his way up the food chain of Adventist college administration when a call came from the U.S. Office of Education in 1964, inviting him to serve as Graduate Research and Program Officer. The family moved to Washington, D.C. and for three years Raymond received an education in the politics of school reform as he directed advanced study consortia with fellows from top tier universities and representatives from the National Education Association and other organizations. Throughout this time, he grew increasingly worried over rising rates of "learning failure and delinquency" and alarmed at the consensus of expert opinion about how to solve such problems. Government and university people were pushing to lower the compulsory school attendance age and advocating early intervention programs such as Head Start. Moore, in contrast, thought that "whenever feasible children should remain longer in the home." He left in 1967, embittered and disillusioned at what he took to be the betrayal of the nation's children by politicians driven more by ideology than by solid research.[22]

About that same time, the Hewitt family donated $750,000 to the Moores. They revived their foundation and set out on an ambitious project to synthesize over 3,000 studies of childhood development to ascertain the impact of early school attendance on children. They found that children were not emotionally, mentally, or physically ready for school until age eight to ten, the very time Adventists had long held that formal education should begin. Their findings were originally reported in the *Phi Delta Kappan*, but the Moores were eager to reach a broader audience. In July of 1972 Raymond and his son Dennis published "The Dangers of Early Schooling" in *Harpers*. The article got such a powerful response that *Reader's Digest* picked it up, publishing an abbreviated version entitled, "When Should Your Child Go to School?" in October. That same month a fuller version of the paper was published in the *Congressional Record*.[23]

All versions of the article made the same basic argument. The Moores presented what they took to be a comprehensive survey of three types of research, all of which concluded against early school attendance. The first group of studies compared children who attended school at young ages with those who did not and found that children who delayed attendance quickly caught up with and surpassed their peers who had been in school longer. The second type of study related to childhood brain formation, finding that neural and sensory-motor development is gradual, which implies that early academic training is premature. Thirdly, the Moores summarized psychological literature finding that early separation of children from their mothers leads to all sorts of emotional and mental problems. The lesson of all this research was clear to the Moores, "schoolmen must realize that there is less value in attempting to substitute for the parent than in helping parents to help themselves and their children." The *Harpers* article, situated as it was in a liberal periodical in the early 1970s, was especially compelling because of its winsome combination of moral outrage at bureaucratic mismanagement and appeal to the commonsense instincts of average parents even as it claimed scholarly rigor and scientific legitimacy.[24]

The response was immediate and profound. The *Harper's* article "elicited one of the largest responses in that magazine's history." Some readers agreed wholeheartedly with the Moores' critique, interpreting it as a progressivist challenge to post *Sputnik*-style drill-and-kill curriculum. Parents especially seemed empowered by the Moores' challenge to professional conventions and pretensions. Educational researchers, on the other hand, were cool in their reception. One researcher whose work had been cited by the Moores objected that her study had been misrepresented. Another found the Moores' use of data selective and reductive. Several letters begged for qualifications and distinctions, most especially between a narrowly academic early education all parties seemed to agree was wrongheaded and a more child-centered approach. One unpublished letter felt that Moore's critique hadn't gone nearly far enough. It was written by John Holt (who himself had had an article in the June issue of *Harpers* advocating deschooling) and was the beginning of a "long, cordial relationship" between the two.[25]

The *Reader's Digest* article garnered even more response, so much so that the magazine commissioned and published Raymond and Dorothy's first book, *Better Late than Early: A New Approach to Your Child's Education*, in 1975. One reviewer quipped that the book "should do for early childhood educators what the *Communist Manifesto* did for Calvin Coolidge." Whatever the professionals thought, the Moores were suddenly in demand. The *Congressional Record* piece and subsequent activism secured for their foundation a $257,000 federal grant. The Moores used this money to amass more research material validating their contention that children should not go to school until ages eight to ten and also to direct research projects at Stanford University, the University of Colorado Medical School, the National Center for Educational Statistics, and Andrews University, all of which came to the same conclusion. In report after report and speech after speech, Raymond Moore repeated the finding that no study exists that clearly points to the desirability of early schooling or daycare for normal children from "reasonably good homes," and that much evidence exists to suggest that keeping children from structured learning environments until ages eight to ten and replacing them with "warm and regular time spent close to parents" was the best predictor of "eventual stability and cognitive maturity."[26]

Then came the Christian tipping point. James Dobson, well-known Evangelical psychologist, author, and radio host, read *Better Late than Early* and was converted to the Moores' perspective. He invited Raymond and then the couple to do a series of shows for his radio program *Focus on the Family* in 1977, which was broadcast over 200 Christian radio stations. "The reaction of our listeners was incredible," Dobson later recalled. "We received more mail in response to those programs than any we had ever aired, with better than 95 percent of the letters being highly favorable." Over the next two decades the Moores would record twenty-one programs with Dobson, for whom homeschooling quickly became a favorite topic. At the beginning, the Moores remained focused largely on the narrower topic of early childhood education. Throughout the 1970s their foundation continued to churn out government reports, textbook chapters, and articles related to the undesirability of early formal training. In 1979 Brigham Young University Press published the Moores' second and most ambitious book summarizing their foundation's latest research on the dire consequences of early schooling, titled *School Can Wait*. This book was the culmination of a research agenda that had begun in the late 1930s with the dissonance Dorothy felt as a young teacher in the California public schools. It remains the greatest scholarly defense of Ellen G. White's views on child-rearing.[27]

By the time *School Can Wait* came out, the Moores had begun to shift their focus. Research papers, government briefings, and textbook production that had preoccupied Raymond in the mid-seventies gave way quickly to a flurry of popular books. Even as he had been overseeing research projects, Moore was increasingly being called before state governments and courts to testify on behalf of homeschooling families. These experiences, along with recurring Dobson appearances, led Moore increasingly to think of homeschooling not simply as an ideal for very young children until they were ready for school but as a viable pedagogy for older children as well. Dorothy too was thinking along these lines, working out the beginnings of what would soon become an ambitious curriculum initiative. In 1981 the couple published *Home Grown Kids*, beyond question their most influential work within the homeschool community. In *Home Grown Kids* the Moores set aside their research-based talking points and laid out a comprehensive child-rearing manual. It was written in an accessible, folksy style well suited to the emerging Christian homeschooling audience, published by a mainstream evangelical press, and plugged heavily by Dobson, who carried it on his mail-order service. The Moores captured perfectly the emerging evangelical consensus on the importance of family values for the Nation's well-being:

> The family was given to us by the same God in whom our country trusts... Nevertheless, we have gone a long, long way toward putting it down and substituting parenting-by-state. Now leading social researchers predict the death of our democratic society within a generation. If we are to retrieve it—and our schools—we would do well to look again to God and the home.

The Moores spoke the language of evangelicalism so well that very few of their readers had any idea that most of what the Moores recommended was lifted straight from the collection of Ellen G. White's domestic writings frequently reissued by Adventist presses, called *The Adventist Home*. Their appearances on Dobson plugging *Home Grown Kids*, as we shall see, proved a seminal moment for many future Christian homeschooling leaders.[28]

By 1982 Raymond Moore had become the most widely known and sought after homeschooling leader in the country. His rhetoric was squarely evangelical, but he eagerly made common cause with the Mormon Freeman Institute, with Roman Catholic Phyllis Schlafly's Eagle Forum, with John Holt, and with anyone else committed to educating children at home. He provided expert testimony in over twenty states, usually *pro bono*, on behalf of homeschoolers in trouble for violating compulsory education statutes and before legislatures considering revising these statutes. He appeared on *Donahue*, *The Oprah Winfrey Show*, *The Today Show*, and many other mainstream venues. In 1983 Dorothy created the "Moore Academy," a tutoring service where counselors worked with families to develop individualized curricula for each child according to his or her interests and abilities. Her "Moore Formula" emphasized remunerative labor, community service, and child-directed study, and it offered families seeking an official school with which to register a legal outlet. Throughout the 1980s and into the 1990s the pair produced a new book every year or two, all emphasizing practical advice, old-fashioned values, and *laissez faire* pedagogy.[29]

Let us briefly summarize what we have covered so far with respect to Holt and the Moores. Though they all came of age in the interwar years and were marked by World War II, both the Moores and Holt grew critical of the postwar consensus about public education. Both began as critics of schooling and only gradually came to see homeschooling as a separate option. Most significantly, both Holt and the Moores enjoyed a pivotal moment when their message was brought to thousands of Americans at just the right time by key media outlets: Holt on *Donahue* in 1978 and the Moores on *Focus on the Family* in 1979, 1982, and many occasions thereafter. These two media events more than anything else put homeschooling on the public agenda. We saw in chapter four how both the Christian right and the secular left were both primed for such a message. The Donahue and Dobson broadcasts brought the message, and they remain the stuff of legend among many homeschoolers today. Finally, both Holt and the Moores came out with their big homeschooling how-to books in 1981: the Moores' *Home Grown Kids* and Holt's *Teach Your Own*. Holt and Raymond Moore quickly found one another, sharing stages at conferences across the country, exchanging information, and even engaging in a friendly debate about the precise number of homeschoolers in the country, with Moore claiming nearly 500,000 by the early 1980s and Holt guessing a far more conservative 10,000. By 1983 the two men were easily the most popular activists for the homeschooling movement. But in less than a decade, their influence had been eclipsed almost to the point that it was no longer remembered by the movement they had shepherded. What happened? To answer that we must begin with another pioneer, one not often noted by homeschooling advocates as a founding father, but whose influence has, in my view, been every bit as profound as that of Holt and the Moores.[30]

Rousas J. Rushdoony

Rousas J. Rushdoony was a second-generation Armenian American. Rushdoony's father had been orphaned at an early age when his father, a priest in the Armenian Orthodox Church, was murdered by Muslim Turks. He was sent to an orphan school run by American Presbyterian missionaries. Thriving there, Rushdoony's father became a Presbyterian and was eventually sent to Edinburgh for study. Returning to Armenia, he and his wife narrowly escaped the Armenian genocide during World War I. They fled to Russia and then on to the United States where Rousas was born in 1916. He attended the University of California, Berkeley, where his thinking first switched "over to English." He earned a B.A. in 1938, a teaching certificate in

1939, and an M.A. in education in 1940. After further study at the Pacific School of Religion, Rushdoony was ordained in the Presbyterian Church, USA (PCUSA), and sent to an Indian reservation in Nevada where he pastored a church for eight and a half years. In 1958 he left the PCUSA and joined the more conservative Orthodox Presbyterian Church.[31]

In the 40s Rushdoony had come under the influence of the Dutch Calvinist philosopher Cornelius Van Til. Van Til argued that first principles, or "presuppositions," must be assumed in an argument, not argued for. Van Til's and Rushdoony's presupposition was that the Bible was God's word, and that there was no neutral ground from which that claim could be debated rationally. Once one assumed its truth, however, a powerful philosophy could be articulated. Rushdoony took this biblical presuppositionalism much further than Van Til ever did. Rushdoony's deep roots in Armenian/Persian modes of thought are readily discernable in his use of the Bible. For him, the Bible should serve roughly the same function in a Christian society that the Koran does in a Muslim one. The Bible is the divinely revealed template for governing every aspect of human life.

Rushdoony's biblicism became the basis for all of his major ideas, two of which have been very influential among some conservative Christians. He is considered by many to be the founder of Christian Reconstructionism, which argues that the Law of God as found in the Bible, both Old and New Testaments, should be the law of nations, especially the United States. Secondly, his "providentialist" interpretation of history has become a powerful challenge to more mainstream Christian historiography. Most Christian historians write history with interpretive parameters that are all but indistinguishable from secular history. But Rushdoony thought of history not as a narrative of human actions but as a revelation of God's sovereign will. And what does God will? That all nations be brought to obedience under the Lordship of Jesus Christ, a philosophy of history commonly called "postmillennialism" because it argues that Christ will return to earth after the millennial kingdom has been fully established by the church. The United States, in Rushdoony's view, has been chosen by God to serve a special role in the unfolding of this Divine plan, as evidenced by the Christian basis of its founding and early history. Rushdoony was one of the first and most effective popularizers of the thesis that America was until very recent times a Christian civilization with Christian laws and politics. Commentator Jeff Sharlet has argued that this providentialist view of history, and not his reconstructionist politics, is Rushdoony's lasting contribution. But for Rushdoony himself, the two ideas were of a single piece. The God of the past and the God of the future is the same God. Recovering a Christian past is but prolegomena for envisioning a Christian future.[32]

Rushdoony's notions that the Bible should be the basis of American law and that the Church's triumph is the purpose of history have had a significant impact on American religious thought. They have been popularized (and watered down) by many commentators. Rushdoony's sharp contrast between a "Biblical" world view and the world view of "secular humanism" has become a staple of conservative Christian cultural analysis, prominent in the works of such best-selling authors as Francis Schaeffer and Tim LaHaye as well as a multitude of B listers. Schaeffer especially was indebted to Rushdoony's analysis and echoed his call for a return to biblical cultural standards. But most of Rushdoony's intellectual heirs, Schaeffer included, have shied away from taking this rhetoric as seriously as Rushdoony did. When Rushdoony called for a return to biblical law, he meant it. If the Old Testament instructs us to stone homosexuals and disobedient children, then this is God's will and the Christian society will do so. Many religious conservatives call for America to turn "back to the Bible," but few mean by that anything so robust as what Rushdoony laid out in his three-volume *Institutes of Biblical Law* (1973–1999). Historian George Marsden calls this less vigorous outlook "soft Reconstructionism," since it calls for "a Bible-based civilization but not in a literal or thoroughgoing way." He notes how odd it is that this orientation has become so common among people who are not postmillennialists like Rushdoony. The vast majority of fundamentalist and evangelical Protestants in America are premillennialists who believe that Christ could return at any time, and that the millennial kingdom cannot be built here and now. These same people, however, often speak "as though they were postmillennialists," devoting "massive efforts to transform American politics and culture for the long run" even as their eschatology suggests that "for the United States there would be no long run."[33]

Rushdoony himself shared in none of this ambivalence or incoherence. Because of his uncompromising and quite alarming politics, Rushdoony's name is not nearly as well known as some of the ideas he originated. Indeed, politicians in his orbit have often felt more than a little sheepish about the association once it is made public by the press. Rushdoony's Chalcedon Foundation, a think-tank he founded in 1965, was for decades bankrolled by reclusive Orange County billionaire Howard Ahmanson, Jr., who is on record as supporting "the total integration of biblical law into our lives." The combination of theocratic beliefs and millions of dollars strikes terror in the hearts of many who fear a Christian usurpation of the democratic political process. Rushdoony and reconstructionism are typically the main characters in progressivist exposés of nefarious plots by the religious right to impose theocracy on America. As such, many Republican candidates who have been offered support by Ahmanson over the years have returned his contributions, fearing that the association would make them look like theocrats. But while many secular Americans fear reconstructionism, some conservatives find Rushdoony and the entire Christian reconstructionist enterprise laughable. Ross Douthat, writing in *First Things*, describes reconstructionism as ranking "somewhere between the Free Mumia movement and the Spartacist Youth League on the totem pole of political influence in America."[34]

Douthat may be right in terms of national politics, but in the homeschooling movement Rushdoony's influence has been direct and powerful. His writings have bequeathed to the conservative wing of the homeschooling movement both a strong sense of opposition between God's law and human laws and a tendency to think of itself as a divinely guided instrument in restoring a Christian America. Many homeschooling families and organizations are every bit as serious about integrating the Bible into public and private life as was Rushdoony, and they see the homeschooling of their children as the first step in the process of raising up what Michael Farris has called the "Joshua generation." "The homeschooling movement will succeed," writes Farris, "when our children, the Joshua Generation, engage wholeheartedly in the battle to take the land."[35]

Rushdoony came only gradually to the homeschooling position, though he had long seen education as a key battleground between the forces of statist idolatry and Christianity. His first work on education, *Intellectual Schizophrenia* (1961), pointed out several internal contradictions in the public school system of the United States, all of which derive from his signature thesis that modernity's elevation of human autonomy is fundamentally at odds with a Christian understanding of the world. Schools reject God but still want students to believe in things like truth, justice, and virtue. They reject the biblical account of creation but still want students to find the environment meaningful. Their commitment to Darwinism leaves human life the product of random mutations, but schools still seem to want students to believe that they possess autonomy and rationality. All of this Rushdoony found breathtakingly incoherent.

In 1963 he published his best known work on education, *The Messianic Character of American Education*. Like its predecessor, it is all critique with only the thinnest of positive programs. Twenty-one of its chapters criticize leading educators, from nineteenth-century reformers James Carter, Horace Mann, and Henry Barnard, to twentieth-century figures G. Stanley Hall, John Dewey, Edward Lee Thorndike, and many others. The final two chapters trace the godlessness of American public education to its rejection of the doctrine of original sin and individual responsibility. The state becomes god and the individual is subsumed into the mass. Finally, at the end of the penultimate chapter, Rushdoony offers the vaguest outlines of a corrective. He calls for a minority uprising:

> The state as yet does not hinder men from establishing and maintaining schools to further their own faith and principles. The future has never been shaped by majorities but rather by dedicated minorities. And free men do not wait for the future; they create it.

In 1963 the dedicated minorities for which Rushdoony was looking were only beginning to surface. As he found them and they found him, Rushdoony became more and more involved in advocacy for both independent Christian schools and homeschooling.[36]

Throughout the 1960s and 1970s, it is clear from Rushdoony's writings and public addresses that the family was becoming an increasing concern for him. His 1965 *Nature of the American System* extended his critique of public schooling beyond his now familiar thesis of a conflict of world views. The battle was also personal: "The concept of 'democratic' or statist education has waged war, not only against Christian faith, but against the family as well." By the seventies he had connected the dots between his critique of the institution and his concern for the family. His 1973 *Institutes of Biblical Law* stressed that "the best and truest educators are parents under God. The greatest school is the family."[37]

Rushdoony's third book on education, *The Philosophy of the Christian Curriculum*, appeared in 1981. Much of it was based on talks he had given to various Christian schools and organizations, and it reflects his growing interest in providing not only a stinging critique of public education but also a robust Christian alternative grounded in the bold initiatives of dedicated minorities. The Christian curriculum will not be just a baptized version of public education but an entirely different animal, built on the Bible and geared toward producing people who have acquired "principled knowledge" to fit them for the purpose of "exercising dominion over the world." Rushdoony works through all of the typical subjects, making his characteristic contrast between humanistic and Christian presuppositions. In mathematics, for example, even so simple a statement as $2 + 2 = 4$ is loaded with assumptions. A true monist, like some of the ancient Greeks or Hindus, would deny the existence of pluralities like 2 and 4, holding that the only true statement is $1 + 1 = 1$. Conversely, without a commitment to metaphysical reality, our notions of 2 and 4 cannot be grounded and are not permanent—they may change tomorrow or yesterday. Only a philosophy that believes both in the unity of truth and the plurality of objects can ground a claim like $2 + 2 = 4$. The Christian Trinity, the notion of one God in three Persons, is the solution. It allows that ultimate reality is both one and many, and hence that mathematics is accurate when it describes many (2s and 4s) even as it assumes that the relationships between them are fixed and eternal. To take another example, Christian study of foreign language is based not on a humanistic desire to "enrich" the student, but serves two purposes. First, study of Greek and Hebrew deepens a student's knowledge of the Bible. Second, study of all modern languages are of value because the Christian is called "to be an imperialist in Christ" everywhere. To convert and rule a people one must know the language they speak.[38]

This robust account of authentically Christian curriculum made Rushdoony a popular figure among many in the independent Christian school and homeschool communities. He was frequently called upon by them to testify in court cases where their legitimacy was being questioned by government officials. Rushdoony's testimony was often taken seriously by juries and judges both because of his education degree and his rhetorical power as a public speaker. Though he died in 2001, Rushdoony's views live on in the Chalcedon Foundation, run by his son Mark. Its monthly publication regularly features homeschooling-themed topics. A recurring theme in these reports is the notion that, as one 1996 article put it, "someone's children are going to inherit the future. It is God's will that it be ours." Family life is the true "training ground of dominion," and homeschooling is the best means, as the March 1998 report notes, for "Raising a Victorious Army for Jesus Christ." One may wonder how representative such views are among the homeschooling population. Are we dealing here with a radical fringe, or does Rushdoony's impact extend beyond the small band of card-carrying Christian reconstructionists?[39]

We have traced thus far the careers of three prominent early leaders: one an atheist, one a moderate Seventh-Day Adventist, and one a very conservative Calvinist. In the early days of the homeschooling movement, these three men and the constituencies for whom they spoke worked together to carve out public space for the practice of homeschooling. Organizations were formed, demonstrations held, legislation challenged or supported, and court cases fought that brought together a wide range of people united in support of the right to teach their children at home. But as the 1980s wore on, this loose coalition of hippies, leftist activists, family values conservatives, and Christian restorationists began to fall apart. The next chapter tells the story of the organizations, magazines, curriculum providers, and leaders of the 1980s and 1990s and how they went from being loosely united to being strongly and bitterly divided.

Chapter Six

The Changing of the Guard, 1983–1998

We must not let a focus on key leaders disguise the truth that the homeschooling movement from its earliest days was thoroughly a grassroots movement. Holt and the Moores were simply catalysts, accelerating trends that were already afoot, converting fence-sitters, and facilitating networks of like-minded families. And networking is what homeschoolers did best. Very quickly in the late 1970s and the 1980s, homeschoolers organized themselves into support groups all over the country. In the early years these groups were usually inclusive, meaning that they accepted all comers regardless of religious affiliation or pedagogical philosophy. Homeschoolers in those days were in a precarious position—misunderstood and held in suspicion by neighbors and family members, distrusted and occasionally persecuted by authorities, confused about what was legal and how to do what they were trying to do. Support groups were a lifeline for many struggling homeschooling mothers: providing sympathetic ears, advice for the daily grind of teaching, and especially expertise regarding how to navigate the educational and legal system. One mother described how a support group "is really the key to being successful.... If one of us is having a horrible day, we can talk each other through it.... It keeps me sane." Another explained how her support group:

> is like an extended family that keeps you on the right track and helps you when you get discouraged. I have gotten a lot of neat ideas from other mothers and I have made several friends and found friends for my children. The most important thing of course is that the group helps me keep my eyes on Jesus.

But therein lay the rub. Could people who were not interested in keeping their eyes on Jesus still be in these groups? Could Christian and secular homeschooled kids be friends?[1]

It did not take long for fault lines to emerge. Conservative Protestants have always made up the majority of homeschooling families, and in the 1980s their ranks grew considerably. In the early 1970s there were perhaps 10,000 to 15,000 homeschooled children in the United States. By the mid-1980s the best scholarly estimates place the number at somewhere between 120,000 and 240,000. Researcher Patricia Lines estimated that by 1990 between 85 and 90 percent of homeschoolers were conservative Christians. Why such a one-sided growth? Homeschooling is nearly impossible without at least one full-time houseparent, and the conservative Protestant celebration of the stay-at-home mom gave it a far larger population of possible recruits than more liberal orientations had. Furthermore, as we saw in chapter four, conservative Protestants had become deeply suspicious of public education by the early 1980s. Many families began homeschooling precisely to escape secular people and ideas, and they were not about to flee from secular liberalism in one place only to embrace it in another. As homeschool leader Scott Somerville noted, many conservative homeschooling mothers "were understandably shocked by the lifestyles of some of their fellow homeschoolers" that they met in the local support group. At the same time, less conservative homeschoolers could feel just as shocked. Lori Challinor, though a Christian, was rejected from her local support group. "I didn't make the grade by their standards," she said. "I wanted a homeschooling support group to provide homeschooling support. I didn't expect the inquisition."[2]

In some locales, people figured out ways to work together. In others, membership was so homogeneous that no fault lines developed. In some others, a majority was able to define the aura of a group, and people of different persuasions took the hint and left. But in many support groups across the country, bitter divisions occurred, resulting often in one support group for conservative Protestants and another (often smaller) for everyone else. The Protestant groups usually required leaders, and sometimes all members, to sign statements of faith, a requirement that scandalized many who found themselves excluded. But such statements did help secure committed and motivated volunteers. As Somerville put it, "a voluntary association exists only as long as there is a shared vision that is strong enough to motivate unpaid volunteers to sacrifice their immediate and personal interest for the common good." Homeschooling is so difficult, and the conveniences of modern life so hard to resist, that only something like "a commitment to a radical 'separation from the world,' with eyes fixed heavenward" can provide many with the inner resources necessary to do it. Exclusion and shunning of those who do not share such commitments is a necessary if ugly concomitant to building strong community bonds. Homeschooling groups that have tried to embrace all views have usually been chronically understaffed, underfunded, and disorganized.[3]

This fissuring of the movement into two factions has been the subject of most of the best scholarly writing on the homeschooling movement. Jane Van Galen, in one of the earliest and best doctoral dissertations on homeschooling, coined the terms "pedagogues" and "ideologues" to connote the two groups. Her terminology was embraced by other researchers, most notably the authors of a widely cited 1992 article in the *American Journal of Education*. For much of the 1990s journalists doing a quick literature review in preparation for a piece on homeschooling would easily find this terminology and use it in their stories. But there is an obvious problem with calling conservative Protestants ideologues and more liberal homeschoolers pedagogues. Both groups were clearly driven by ideological commitments, and both certainly employed a wide range of pedagogies. More recently, sociologist Mitchell Stevens has suggested calling the conservative Protestants "believers" and other homeschoolers "inclusives." This terminology happily avoids the connotation that only liberal homeschoolers had pedagogical motivations and only conservatives had ideologies, but it too runs the risk of implying that only conservative protestant homeschoolers are religious believers or that all believers are separatistic.[4]

I would like instead to use a distinction found in Christian debates over who may receive communion in a local church as a metaphor to capture the division. Many Christian groups practice "closed communion," meaning that only members of their particular group may receive communion during worship in a local church. Other groups practice "open communion" where all present may receive regardless of denominational affiliation, so long as certain minimal requirements are met. It should be noted here that the term "communion" itself has often been a source of division among Christians. Many Protestants from Free Church traditions reject the term entirely, preferring instead phrases like the "Lord's Supper" or "table fellowship" to designate the shared bread and cup. Many of these Christians, well represented among the homeschooling community, find the term unscriptural and too laced with medieval theological baggage. In using it, I by no means wish to imply that such Christians use the term themselves—it is strictly a metaphor to help illustrate how different groups of homeschoolers have dealt with the issue of religious pluralism.

The distinction between closed and open communion is really a continuum, as churches "fence the table" to varying degrees. Some require strict adherence to denominational particularities and/or rigorous preparation like fasting and confession. Others require Trinitarian baptism or evidence of conversion or some other broadly Christian commitment. The most liberal churches allow even non-Christians to receive. In much the same way, homeschoolers since the mid-1980s have typically fallen somewhere along the continuum of an open communion or a closed communion approach to other homeschoolers. For the remainder of the book I will refer to homeschoolers who excluded others from their groups, conferences, and publications as "closed communion homeschoolers" and those who did not take this approach as "open communion homeschoolers," though we must always bear in mind that both tendencies have come in varying degrees. Many "closed communion" homeschoolers, having separated themselves from secularists, Mormons, Jews, Catholics, and Pagans, still found themselves confronted with intra-Protestant doctrinal differences over such issues as predestination, infant baptism, eschatology, charismatic gifts, divorce and remarriage, female head-covering, dating, rock music, and many other issues over which biblicist Protestants disagree. Some separated still further, while others accepted differences at this level. At the same time, many open communion groups developed an exclusivity of their own, as often the only thing uniting their diverse clientele was an antipathy to closed communion Protestants.[5]

Local fissures between closed communion and open communion homeschoolers were dramatically replicated (and fostered) at the national level. In hindsight, it seemed bound to happen. The two public faces of homeschooling in America, John Holt and Raymond Moore, were in many ways out of step with the overwhelming majority of their constituents. Holt had no religious commitments, never married or had children, and seemed for much of his career to think of the family as a hindrance to individual self-actualization. For closed communion types, this was simply unthinkable. To this day many accounts of the history of homeschooling written by closed communion partisans do not even mention Holt or the entire left wing of the movement. Moore, though he seldom discussed the matter, was Seventh-Day Adventist, a group many evangelicals think of as a cult. In addition, his gentle ideas about early education worked against the grain of the emerging conservative consensus that tended to favor rigorous doses of phonics, firm discipline, worksheets, and memorization. Moore's emphasis on health, nature, and child-directed study sounded like new-age liberal relativism to some. Finally, there was a generational gap between these men and the movement's rank and file. In 1983, a pivotal year in the history of homeschooling, Holt was sixty and Moore sixty-eight. In two years Holt would be dead from cancer and Moore would be increasingly on the defensive as a younger and more aggressive group of baby-boomer Christian leaders emerged.

Christian Liberty Academy

The seeds of closed communion unrest with their spokespersons were planted well before the baby-boomer takeover. Rushdoony never had a breakout moment or a best-seller like Moore and Holt, but he influenced key people who quietly built lasting organizations. One of the most important was Paul Lindstrom, a public school teacher-turned Calvinist pastor and homeschool activist. Lindstrom founded the Church of Christian Liberty in 1965, and it grew in the late sixties into "a self-consciously Reformed and Reconstructionist hotbed of Christian activism" in Arlington Heights, IL. Lindstrom began his conversion to homeschooling in 1966 when he discovered phonics. By 1967 he had created a correspondence school modeled after the venerable Calvert program but with an explicitly Christian curriculum. He called his organization the Christian Liberty Academy Satellite Schools (CLASS) and urged parents, in his own words, "to remove their K-12 children from public schools and, with or without local approval, simply teach them at home." Lindstrom was "permanently shaped" by Rushdoony's message and life, and the two quickly established a partnership. The CLASS curriculum emphasized "the Reformed world and life view, the only theological position that consistently leads Christians into meaningful and victorious interaction with the culture." Students were assigned many explicitly reconstructionist texts, most notably *Biblical Economics* by Rushdoony's son-in-law Gary North, and every subject was taught in accord with Rushdoony's *Philosophy of the Christian Curriculum*, with the Bible serving as both textbook and theoretical guide.[6]

As word of Lindstrom's program spread, Christian homeschoolers around the country began taking advantage of this inexpensive and pervasively Christian curriculum. Over the next decade nine of the families using CLASS curriculum were taken to court by various school districts. Lindstrom hired a number of lawyers known to be sympathetic to Christian schooling, some of whom would later go on to be key homeschooling activists themselves: John Whitehead (who would later found the Rutherford Institute), Michael Farris (future founder of the Home School Legal Defense Association and Patrick Henry College), William B. Ball (lead counsel for many landmark religious freedom cases), David Gibbs (founder of the Christian Law Association), and others. Lindstrom's organization won seven of the cases, but more importantly, the press coverage given to them spread word of Lindstrom's correspondence program even further.[7]

By 1983, there were 6,000 students enrolled in Christian Liberty Academy, and several hundred were staying with it through 12th grade. A growth explosion occurred in the next two years, when enrollment skyrocketed to 21,000. By the mid-1980s CLASS had become the largest correspondence K-12 program in the country, and the numbers kept going up. By 1998, the high point, there were 35,000 children enrolled and over 800 high school graduates. Though the numbers have declined steadily since then, the thirty-year dominance of CLASS as provider of Christian homeschooling curricula has had a lasting impact on the movement's political and theological orientation. Chris Klicka of the Home School Legal Defense Association went so far as to say that "without Pastor Lindstrom and his ministry, I don't think the homeschool movement would have gotten off the ground. I think they were that integral in providing the nuts and bolts curricula, and also the spirit to hold on." That spirit was the militant reconstructionism of Rousas Rushdoony, which CLASS has now passed on to over 100,000 American families. "Our continuing mission," wrote Church of Christian Liberty Associate Pastor Quentin Johnson in 1998, "is to provide inexpensive, quality education to those who desire to prepare themselves and their children for their part in taking godly dominion in every aspect of life and thought." CLASS remains committed to this vision today. As their Web site proclaims:

> CLASS desires to train Christian warriors and leaders who will go forth in the power of the Holy Spirit to win decisive victories for the honor, glory, and kingdom of Christ.... Christian education is a part of God's purpose to put everything in creation under the feet of Christ. Christ is now reigning, but not everyone has bowed the knee to His authority. Not every area of life and thought has been made subject to Him. Christian education serves as part of His purpose to subdue people from all nations to the glory of God.[8]

Lindstrom's program popularized Rushdoony's approach to the Bible, education, and political engagement among thousands of homeschooling families. In so doing it laid the groundwork for the meteoric rise of several individuals and organizations that emerged in the early 1980s and very quickly took control of the homeschooling movement, giving it a distinctively dominionist flavor. Ironically, many of these second-wave Christian leaders were first exposed to the idea of homeschooling through Raymond Moore's Dobson shows. Many of them used Moore's influence at first to gain a following, but as soon as they could stand on their own, they quickly marginalized him. In the text that follows, we will look at some of the major closed communion leaders, all of whom emerged at about the same time in the early 1980s and quickly came to dominate the homeschooling movement internally and to serve as its public face to outsiders. We will look first at seminar leaders, then curriculum providers and gatekeepers, and finally the lawyers who consolidated the Christianization of the movement.

The Seminars and Conferences

In 1968, fifteen-year-old Gregg Harris quit high school, ran away from his rural Ohio home, and headed to Laguna Beach to be a hippie. By the 1970s the disillusioned young man had been caught up in the Jesus movement after being evangelized by Calvary Chapel preachers. The prodigal returned to Dayton and began attending the First Baptist Church with his family, where he met his wife Sono. Eager to serve God in ministry, Harris enrolled in the fledgling Centerville Bible College near Dayton, where, after a year of study, he was ordained and sent out to Harlingen, TX, to plant a church that he dubbed "Full Gospel Fellowship." In Harlingen, Harris began to discover the limits of the hippie Christianity in which he was immersed, and he grew increasingly embarrassed by his own lack of education and spiritual preparation for ministry. He disbanded the church, returned to Ohio, and began taking education classes at the University of Dayton while serving as senior pastor at the Grace Fellowship Church in town. In 1980 the Harrises placed their first child, five-year-old Joshua, in preschool at Dayton Christian Elementary School. They quickly noticed a change in their son. At home "he'd suddenly burst into tears for no apparent reason." But Josh had a reason. He was afraid of school because some kids there were bullying him: "After being pushed down and intimidated a few times he began to hide inside himself." Gregg was concluding from his education classes that schools met the needs of educational professionals, not children. So they pulled Josh from school and began to educate him at home. Bud Shindler, Dayton Christian's director, was willing to work with the family and together they crafted an extension program whereby Josh was permitted to homeschool legally under the auspices of Dayton Christian.[9]

After two years at the University of Dayton, Harris transferred to Wright State University to study conference planning and management. What he really wanted, however, was a mentor. Harris was becoming convinced that the Church needed to do a better job preparing leaders by having them apprentice with established, mature Christian men. He was also developing a vision for raising up godly families in the United States. Around this time Harris met Bruce Wilkinson, founder of Walk Thru the Bible Ministries (and future author of the runaway best-seller *The Prayer of Jabez*). Wilkinson was one of a large group of successful Evangelical entrepreneurs who were making a good living touring the country offering seminars to various Christian groups. Harris had thought of joining this trend and was working up a series of talks on Christian homeschooling. Wilkinson became something of a model for Harris. "He was like Solomon to me," Harris later recalled, "Shaking his hand...listening to his counsel and following through with his advice...set a number of things in motion in my life. Dr. Wilkinson is, in a sense, my hero." In 1981 Harris began his own ministry, offering a two-day Christian Life Seminar and a Home Schooling Workshop covering the benefits of homeschooling, some how-to techniques, and a broader theology of family that included strong doses of patriarchal leadership, firm discipline, holy living, and home-based economy. At first he met with little success. In 1982, to draw a bigger crowd, Harris brought Raymond Moore to the University of Dayton for a conference. Moore was impressed with Harris and offered him a job. In 1983, still eighteen credits shy of earning his degree, the Harris family relocated to Washington to work for Moore. Harris stayed only three months, and shortly thereafter his own speaking career took off. What happened?[10]

What happened is hard to discern. To date, movement insiders and scholars chronicling this story have tiptoed around it. The only public source is an unpublished "white paper" Raymond Moore wrote in 1994 that has been widely circulated among open communion homeschoolers and can still be found on many Web sites. By the time Moore wrote the paper he was struggling with senility, and the rambling and repetitive text reveals as much. But if one reads with patience, the outlines of what Moore thinks happened become clear. As Moore tells the story, Harris secured the job, his first "regular" job, only to subvert what Moore was doing. While the Moores were at church on Saturdays after the Adventist fashion, Harris was in the office photocopying Moore's mailing lists and contact information. While answering the phone when Moore was out of town, Harris would tell leaders of state homeschooling organizations and others requesting Moore's presence as a speaker that Moore no longer was able to come, but that he was sending Harris as a replacement. At first Moore had no idea what was going on, but repeated calls from state organizers asking why he was reneging on speaking commitments alerted him to the situation. So Harris was fired.[11]

Harris has never published a rebuttal to these charges, but over the telephone he gave me his version of the events. Impressed with Moore's advocacy for homeschooling, Harris thought perhaps Moore could be the mentor for whom he was searching. When Moore offered Harris the job, Harris believed that he would be given opportunities to develop his speaking ministry under Moore's tutelage, a ministry that would eventually grow to the point where he could sustain himself. Shortly after he had come on board with Moore he gave one talk at a Moore conference in Portland. Michael Farris had also been invited by the Moores to speak. Farris' and Harris' sessions ended up being the conference highlights, and tapes of their sessions sold far more copies than sessions offered by the Moores. This bothered Moore and he changed Harris' job description from conference speaker to conference organizer and had him stay home for the next big conference.

Harris, however, by his own admission, was not a good administrator. His wife Sono was, and Moore suggested that she place Josh and their one-year-old son Joel in childcare so that she could help run the organization. This suggestion only added to Harris' frustration over his recent demotion from speaking, so he quit. Harris acknowledges photocopying mailing lists, but he maintains that this was simply part of his job as conference organizer. As for stealing conferences, Harris asserts that while he was on staff Moore gave him explicit instructions only to accept speaking engagements that would guarantee an audience of a certain size in major population centers. Smaller venues Harris was instructed to turn down, which he did. After he left Moore, however, Harris contacted some of these smaller groups and offered to host seminars for them himself.

However the original contacts were obtained and made, Harris quickly developed a strong following among the large portion of homeschoolers who were conservative Christians. All around the country, in church basements, gymnasiums, and anywhere he could draw a crowd, Harris would offer his seminars, always following the same basic marketing strategy. An initial seminar was planned, with advertising being sent out to homeschoolers on Moore's mailing list and existing (often open communion) support groups. Fliers were posted in such conspicuous places as supermarket bulletin boards, local churches, and so forth. At the conference Harris would present his material, which included a strong message that Christian homeschoolers should not be in support groups with other homeschoolers who were not Christian. He urged Christian support groups to be founded on "a statement of faith, which should be affirmed by any potential group leader" and crafted so as to be "narrow enough to exclude people from a nonevangelical framework or with abhorrent opinions."[12]

Typically, his Home Schooling Workshop would "kick off the establishment of state Christian Home Education Associations and metropolitan support groups." Thirty-five statewide Christian homeschooling associations were founded in this way, sometimes as alternatives to preexisting open communion state groups, sometimes taking over the earlier group altogether, and sometimes building from scratch. These groups would then "serve as our annual workshop hosts," thus guaranteeing a perpetual audience and source of income for Harris. It was a brilliant market strategy, and it succeeded. By 1988 Harris claimed to have presented his seminars to over 18,000 families. By 1995 he claimed over 95,000, with individual conferences regularly drawing 1500 people. In addition, large numbers of people heard him deliver portions of his message at other homeschooling venues or on videotape. By the early 1990s Gregg Harris had become one of the most widely known Christian homeschooling leaders, with a smoothly functioning administrative model employing six people to keep networks of supporters in constant motion preparing for his annual conferences.[13]

The rapid growth of closed communion homeschooling groups led to a demand for more people like Gregg Harris to serve as keynote speakers at conventions. As such, many entrepreneurial homeschoolers adopted Harris' strategy and began making the rounds at the various conventions offering their particular slant on homeschooling. Some made their name through political activism and inspired their audience with war stories of God's victory over statist persecution. Some became curriculum experts and expounded the virtues of various approaches to home learning. Some developed messages that dealt with difficulties commonly faced by homeschoolers: feelings of isolation, burnout, or incompetence; frustration at husbands who weren't pulling their weight, concern about teaching older children;

and so forth. Some expanded homeschooling into an entire world view, advocating such things as home birth, house church, and home-based business. A number of these speakers have honed their messages over the decades and continue to make appearances, while many others exploded onto the scene with an idea that was all the rage for a cycle or two and then faded. I cannot detail here all of these trends and personalities. There is one figure, however, who has always existed outside of the world of the Christian homeschooling convention but whose impact on the movement has nevertheless been so profound that he cannot be overlooked: pioneering Christian seminar leader Bill Gothard.

In 1957 and 1961, respectively, Gothard earned his B.A. and M.A. from Wheaton College, the nation's leading Evangelical school, after spending fifteen years ministering to gang members in the Chicago area. Out of this experience, Gothard developed the material that would become the "Institute in Basic Youth Conflicts," premiered in 1964 as a credit-bearing course at Wheaton. Basic Conflicts laid out seven "universal principles of life" that, if followed, would lead to true success in life. "Every problem in life," Gothard would intone for the next forty-five years, "can be traced to seven non-optional principles found in the Bible. Every person, regardless of culture, background, religion, education, or social status, must follow these principles or experience the consequences of violating them." Underlying Gothard's principles is an extensive focus on submission to divinely instituted authorities, especially children to parents and wives to husbands.[14]

In response to the student unrest and youth protest of the late 1960s, Gothard's seminar became increasingly popular among conservative Americans who felt their society was tipping toward anarchy. By 1973 he was taking his seminar on the road, offering a weeklong, thirty-hour experience in cities around the country. By the late 1970s Gothard's seminars were being attended by 10,000 to 20,000 people every time they were offered. They peaked in the early 1980s, when 300,000 people were attending them annually, but have steadily declined since then. In the 1980s and 1990s Gothard created several organizations that expanded on his Basic Youth Conflict seminars, all of them run from his Institute in Basic Life Principles headquarters in Oak Brook, IL. Along with a medical facility, a law school, a male-only martial training school in Big Sandy, TX, and a secularized "CharacterFirst!" moral education package that has been used by hundreds of public schools, Gothard runs the Advanced Training Institute (ATI), a correspondence homeschooling program founded in 1984 that quickly gained enrollments topping 10,000 a year. Families enrolling their children in ATI are frequently alumni not only of Gothard's Basic Conflict seminar but the more rigorous "Advanced Seminar." After mastering his principles, families fill out an extensive application asking very intimate details about their personal lives and submit themselves to intensive training to ensure their conformity to Gothard's standards.[15]

Gothard has long been a very controversial figure in the conservative Protestant world. Many dispensationalists (who believe that Old Testament law was not intended for the current Church Age of history) criticize his approach to the Bible, especially his literal adoption of Old Testament guidelines. Many more moderate evangelicals find his teachings excessively legalistic and divisive. Cult-watchers worry over his authoritarianism and secrecy and swap rumors of sexual indiscretions (Gothard never married). However, thousands of committed homeschoolers have had their lives transformed by Gothard's message of submission to godly authority, courtship over dating, anti-contraceptive "open embrace" sexuality, home birth, resistance to the evils of rock music and minor chords, debt-free living, modest dress, and much more, all packaged in a self-help style that promises great results if listeners follow his principles. Gothard did not become a homeschooling player until the mid-1980s, but once he did, he has succeeded as much as anyone else in tying homeschooling to the broader network of countercultural Christian ideas with which it is still associated in the minds of many. If the public stereotype of the homeschooling family is that of the firm but gentle patriarch, the Titus 2 mom shrouded in a loose-fitting jumper and headcovering, the quiver-full of obedient stairstep children dressed in matching homespun, we have Bill Gothard to thank as much as anyone.[16]

The Curriculum Providers and Gatekeepers

As we have seen, Christian Liberty Academy's correspondence program was throughout the 1970s and 1980s the dominant player in Christian home school curricula. In their early years they provided families with an eclectic assortment of nineteenth-century schoolbooks and the Accelerated Christian Education (ACE) curriculum, created in 1970 by Dr. and Mrs. Donald R. Howard for the fundamentalist school they opened that year in Texas. The curriculum promised to teach children "from God's perspective," celebrating the virtues of

Christian America and free enterprise as it decried communism, socialism, the United Nations, and secularism. ACE grew to be a popular curriculum for thousands of the Christian schools created in the 1970s and 1980s: by 1976 there were 1,450 schools using it. By 1987 around 5,000 were. Christian Liberty Academy was by far ACE's most valuable customer, sending out Dr. Howard's curriculum to thousands of homeschoolers across the country. But the ACE curriculum was a frequent target of criticism from many sides: the mainstream press saw it as emblematic of a paranoid and conspiratorial fundamentalist world view at odds with pluralistic America; many educators decried its drill-and-kill pedagogy; and, most importantly, many homeschooling families chafed against its rigidity even if they endorsed its political vision. What CLASS needed was another curriculum that shared the religious vision of ACE but was more family-friendly in implementation. They found what they were looking for in Bob Jones University Press and A Beka Book.[17]

Bob Jones University Press began with a vision similar to that of the Howards. In 1974 it published the first of many textbooks intended for the growing Christian school market. Like ACE, it sought not to repackage secular textbooks with a few Bible verses thrown in for good measure but to create an authentically Christian curriculum from the ground up. Unlike ACE, however, Bob Jones had the benefit of an entire school of education and contacts with numerous writers and researchers who could work collectively. The result was a coherent, literate curriculum that has proven popular ever since. A Beka Book began in the same way. It was an offshoot of Pensacola Christian Academy, founded in 1954 by Arlin and Beka Horton (hence A Beka). The Hortons developed their curricula in the context of the Sputnik era reaction against progressivism to stress phonics, memorization of facts, and celebration of American patriotism. By the 1970s they were offering their curricula to other Christian schools, and when the homeschooling market opened up, they too were there to capitalize on it, beginning a correspondence wing (A Beka) in 1975. By 1985 Pensacola Christian Academy was enrolling 1,870 students a year over and above the students they reached through CLASS. By 1998 they had over 23,000 students enrolled in the correspondence school and 225,000 more families purchasing their books independently.[18]

Bob Jones and A Beka became widely known among homeschoolers in the 1980s largely through Christian Liberty Academy, but as more and more families sought to homeschool independently, both publishers developed networks whereby the middle man could be cut out and they could deal with customers directly. The most effective of these was the homeschooling convention. In the late 1970s and early 1980s, these conventions tended to be open communion and quite informal, but as Harris and others formed closed communion organizations in state after state, many of these conventions quickly took on a distinctly Christian cast and an elaborate organizational infrastructure. Convention attendees would leave their children behind in order to hear speakers, attend workshops, network with other homeschoolers from their region, and buy materials for the upcoming year. A Beka and Bob Jones curricula were always featured prominently at these conventions.

Given the dynamic nature of this movement and the fact that so many of its practitioners were inclined toward cottage industries and home-based business, very quickly dozens of competitors emerged with their own products. Some, such as Christian Light Education, Konos, Alpha Omega, Rod and Staff, Weaver Curriculum Series, Tapestry of Grace, and Sonlight, were complete curricula school-in-a-box style (many could also be purchased a-la-carte). Others, such as Apologia, Considering God's Creation, Answers in Genesis, How Great Thou Art, and Diana Waring, offered specialized products in specific subjects, especially hot-button Christian subjects like creationism and providential history. Pedagogical approaches ranged from reproducing the traditional classroom at home, to the popular unit-study approach, to more self-directed styles. Some were worksheet heavy or required extensive parental participation and evaluation. Others were more literature-rich or student-led. Many of these products made every effort to draw out the Christian implications of their subject and to have the Bible as interpreted by American fundamentalism permeate the curriculum. Some took a more moderate stance here as well. Very quickly, such a wide and diverse array of options became available that the new homeschooler could understandingly become quite confused and frustrated over the prospect of choosing a curriculum. By the 1990s conventions in many states would have seventy, eighty, or more vendors hawking products or niche ministries. By 2000 the largest state conventions regularly drew over 100 exhibitors.[19]

Here too homeschool entrepreneurs emerged to offer their services, in two forms. First, several veteran homeschoolers became curriculum experts, producing frequently updated meta-analyses of the proliferating options. The two most influential reviewers were Mary Pride and Cathy Duffy. Pride converted to Christianity in 1977 and attended a Reformed Presbyterian Church where she became conversant with the mental landscape and literature of biblical presuppositionalism. Soon thereafter she began homeschooling her children (nine in all). In 1985 she published her first book, *The Way Home: Beyond Feminism, Back to Reality*, an extended meditation on Titus 2:3–5 that included a substantial section praising homeschooling. Noting that there were few how-to books on the market, she published her *Big Book of Home Learning* in 1986. It has been in print ever since, constantly revised and expanded (the 1991 version ran to four volumes). By 1996 its various incarnations had sold a quarter of a million copies. Though Pride covers nearly every topic imaginable in this work, it is most famous for its detailed curriculum reviews. Hundreds of thousands of homeschoolers have listened with care to her evaluations, charmed by her unique combination of New Yorker spunk and conservative stance on marriage and family matters. Pride's rhetoric is that of the leftist counterculture, but her social agenda is as traditionalist as it comes. She rejects birth control, remarriage after divorce, female employment outside the home, and movies and television, while endorsing quiver-full families, home church and business, patriarchal headship, and simple country living. And she does it all with sharp wit and catchy *bon mots* (Planned Parenthood she calls "Planned Barrenhood." Family planning is "family banning.") The "queen of the homeschooling movement" has built a mini-media empire including her magazine *Practical Homeschooling* (Pride has claimed a circulation of over 100,000 for a decade), sprawling Web site, and steady stream of how-to books.[20]

Though not so well known as Pride, Cathy Duffy's *Christian Home Educator's Curriculum Manual* runs a close second in terms of influence on the choices Christian homeschoolers have made regarding curriculum. Duffy began looking into homeschooling in 1981 after her sister passed on Moore's *Home Grown Kids* to her and she had several serendipitous encounters with homeschooling families in California, leading to a conviction that "God was directing our family" that way. By 1984 she was a veteran and began self-publishing her *Curriculum Manual*. Word got around and what began as a "slim, comb-bound" booklet soon expanded into a two volume work. Thousands of readers have found in Duffy's detailed, clever, and candid reviews both a charming authorial voice and a reliable guide through the increasingly complex labyrinth of homeschooling curricula. Her self-published guide has sold over 100,000 copies, largely through word-of-mouth advertising. In 2005 a new version was published by Broadman and Holman as *100 Top Picks for Homeschool Curriculum*. Duffy maintains a popular Web site as well. Several other guides by such authors as Theodore Wade, Mary Hood, Cheryl Gorder, Don Hubb, and Donn Reed have helped thousands of homeschooling families in their search for reliable information, but none have had such staying power or influence as those of Pride and Duffy.[21]

The second mechanism for corralling and directing the explosive growth of Christian homeschooling was the periodical. Until the mid-1980s there were not very many choices available for Christian homeschoolers. As late as 1985 Mary Pride could write, "I strongly recommend subscribing to John Holt's newsletter/magazine *Growing Without Schooling (GWS)*. Mr. Holt is not a Christian, but that shouldn't stop you from enjoying *GWS*, the best home schooling resource around." She deferred to Holt as well for curricular recommendations, "Believe it or not, Holt Associates...can tell you anything you need to know about any of these subjects. There are pamphlets, book lists, curricula lists, reviews of materials, and so on, enough to stagger the imagination." But with Holt's passing, a new generation of explicitly Christian periodicals emerged as alternatives to *Growing Without Schooling*. These publications were nearly always the product of enterprising homeschooling families using desktop publishing technology to launch what they hoped would be a successful home business. Since the homeschoolers most attracted to the home business strategy tended also to embrace a caffeinated version of what David Brooks has called the "natalist" outlook (stressing such topics as home birth, house church, courtship, quiver-full families, modest dress, and rural values), this perspective has perhaps been overrepresented among homeschooling periodicals, leading to a perception that all Christian homeschoolers have at least eight children, watch no television, and eat only homegrown food. Mary Pride's *HELP for Growing Families* and Cheryl Lindsey's *Gentle Spirit*, both begun in 1989, embodied this hyper-natalist sensibility. We will return to their subsequent history shortly. Other lesser-known magazines like *Patriarch*, *Quit You Like Men*, *Above Rubies*, Josh Harris' *New Attitude*, and others won smaller followings and propagated the full-throttled natalist vision or some aspect thereof. Families adopting this outlook often struggled to find acceptance in local churches. Such families often found that their local church worked against the very things they were trying to accomplish by segregating children by age in Sunday School classes and youth groups, by having children leave collective worship for "children's church," and by replacing the authority of the father with the authority of credentialed experts. Many of these families ended up leaving established churches to form small house-based churches of the like-minded.[22]

Less culturally radical magazines emerged as well, some of which found limited success. *Homeschooling Today* was started in 1992 by Steve and Kara Murphy and has proven to be one of the most enduring. The most widely circulated Christian magazine has been Pride's *Practical Homeschooling*, which succeeded where her *HELP for Growing Families* did not. The first, and historically most significant, of the successful Christian periodicals to reach a national audience was *The Teaching Home: A Christian Magazine for Home Educators*, started in 1980 by Sue Welch and her family. Welch lived only a few miles from Harris in Portland, Oregon. Her publication, "produced in a converted garage on a relatively low-end desktop publishing piece of hardware," worked in tandem with Harris' seminars to build a national closed communion Christian homeschool movement. Welch's innovation was to staple into each issue of the magazine a specialized insert targeted to readers in each state. At first many of these inserts were newsletters from the open communion statewide organizations that had been created by Holt, Moore, and others. But as Harris' seminars kicked off closed communion organizations, Welch gradually replaced the open communion inserts with those of closed communion groups. Welch's publication also included extensive information about various support groups and conventions meeting around the country, making it a valuable networking tool. Again, in the early years she included contact information for conventions and meetings of all groups, but as closed communion organizations grew stronger, she stopped listing open communion organizations and meetings. Throughout its history, *The Teaching Home* placed special stress on Harris' Christian Life Workshop tours, and Harris was a regular contributor to the magazine. The cross-marketing between Welch and Harris paid off, and by 1994 *The Teaching Home* had over 37,000 subscribers.[23]

Another notable aspect of *The Teaching Home* was its extensive "Legal News" section. This portion of the magazine was written by the staff of the Home School Legal Defense Association, an organization of lawyers that in the late 1980s and 1990s became the leading national voice for homeschooling.[24]

The Lawyers

Rousas Rushdoony believed that "religions that fail to dominate and control education and law quickly become fading relics of the past." He thus spent nearly as much time raising up a Christian army of lawyers as he did on Christian schools and homeschools. One of his most successful protégés was John Wayne Whitehead. Whitehead did not begin as a Christian Reconstructionist. He had been a leftist radical after college and in law school during the late 1960s and early 1970s, attending war protests, growing out his hair, smoking weed, and parroting Marxist talking-points. He also had a soft spot for science fiction, and when he saw that Hal Lindsey's *Late Great Planet Earth* had sold eight million copies, he bought it, mistaking it for a genre piece. "It spooked me right into heaven," Whitehead said.[25]

Upon his conversion, Whitehead quit his law practice and drove with his wife Carol to Lindsey's Light and Power House Seminary in Los Angeles. While studying to be a pastor, he was approached on several occasions by fellow Christians, who, having learned of his lawyer background, shared with him stories of persecution at work and in school settings. Though many fellow Christians told him that it was unbiblical to be involved in secular courts, Whitehead couldn't separate Christianity from politics. "I became a Christian from being a Marxist, so I still believed the one central idea that Christianity has this political thing tied to it." Whitehead dreamed of beginning a law firm that would defend Christians for free, funded by private donations. He found an ally in Rousas Rushdoony. In 1977 Whitehead published his first book, *The Separation Illusion: A Lawyer Examines the First Amendment*. The book was outlined and introduced by Rushdoony himself, and it argued strongly that Christians must engage the culture in the legal arena lest their absence cede it to secularists. In 1979 the Whiteheads moved back to the east coast and set up a law practice. John was especially enamored with religious liberty cases, and he frequently worked *pro bono*—in one case where he successfully defended a homeschooling family, Whitehead spent $25,000 of his own money. Such commitment and results impressed potential backers and won Whitehead a prominent place in the emerging pantheon of New Right activism. His second book, *The Second American Revolution*, though it was essentially a rehash of *The Separation Illusion*, was released to an evangelical audience that was now politically engaged. Dobson interviewed Whitehead and carried *The Second American Revolution*. It sold over 100,000 copies. What in 1977 had seemed an impossible dream became a reality in 1982. Rushdoony's Chalcedon foundation and Howard Ahmanson provided much of the seed money for Whitehead to realize his vision of a Christian legal society, and the Rutherford Institute was born, with Rushdoony, Ahmanson, and Francis Schaeffer all on the board of directors.[26]

Throughout the 1980s the Rutherford Institute took many homeschooling cases, always for free. Homeschooling appealed to Whitehead both on civil liberties grounds and because it was on the vanguard of what he took to be a restoration of Christian America. His willingness to take on these cases, along with that of a number of other Christian lawyers, met a real need in the homeschooling community, for more established civil liberties groups like the American Civil Liberties Union were for the most part not interested in the homeschooling issue. Nearly as important as his organization's work on behalf of homeschoolers is the book he cowrote with Wendell Bird, *Home Education and Constitutional Liberties*, first published in 1984 but updated and expanded in 1993 as *Home Education: Rights and Reasons*. This book, though now dated, remains the most rigorous summary of homeschooling law and jurisprudence ever written and has had a powerful impact on the movement. In his writings, lectures, legal work, and mentorship of young Christian lawyers, Whitehead laid the groundwork for a Christian legal activism that would be built on by another organization founded in the early 1980s, the Home School Legal Defense Association.[27]

Michael Farris had always wanted to be a lawyer. His father, a public school principal and "fervent Baptist," would often urge him to get a law degree so he could defend public schools against the American Civil Liberties Union (ACLU). Farris did just that, receiving his J.D. from Gonzaga University School of Law in 1976. By then he had determined that the public schools were no longer worth defending. Schools now embraced the same libertine ethic and antipathy toward traditional Christianity as did the ACLU. Full of righteous zeal and indignation, Farris began his practice in Washington prosecuting abortion clinics and pornographers. His compelling stage presence and aggressive rhetoric got him noticed by Jerry Falwell's Moral Majority movement. He became executive director of the Washington branch of Moral Majority in 1979, which he soon renamed the Bill of Rights Legal Foundation, intending it to be a Christian counterweight to the ACLU. Around the same time he also became chief legal council for Beverley LaHaye's Concerned Women for America where, among other initiatives, he filed a successful federal suit to stop the push to extend the deadline for state ratification of the Equal Rights Amendment. Many of Farris' other suits were not so successful, however, and would come back to haunt him in the 1990s when he ran for political office. Farris sued the Washington secretary of state to stop the state lottery that had been passed by the legislature; a Washington school district for including Gordon Park's

The Learning Tree in its curriculum; and most famously, the Hawkins County School District in Tennessee which would not allow the children of Robert Mozert an alternative to the school reader he and other parents found offensive. Stories to which the family took offense included Rumplestiltskin, Cinderella, and the Wizard of Oz. (Farris' political enemies would later make much of this.) In both the *Learning Tree* and *Mozert v. Hawkins* examples, Farris argued, unsuccessfully, that such books unconstitutionally established the religion of secular humanism in public schools.[28]

While his career as one of the go-to lawyers among burgeoning New Christian Right organizations was taking off, Farris' family was growing even faster. Michael and Vicki Farris were married in 1971 and began having children after he completed law school. Over the next twenty-three years they would have ten children altogether. By 1982 their eldest daughter Christy was in second grade. Farris was increasingly involved in litigation against schools, and he and Vicki were at a loss over what to do with their own growing brood. Vicki heard the Moores on Dobson and wanted to try homeschooling, but she worried that her husband wouldn't approve. So she prayed in secret that God might lead Mike to the same conclusion. In the fall of 1982 Farris met Raymond Moore in Utah when both appeared as guests on Tim and Beverly LaHaye's radio program. Moore convinced Farris to homeschool, and he went home to share the news with his ecstatic wife. Later that year Farris attended a Raymond Moore conference in Dallas where he met many like-minded people, especially Kirk and Beverly McCord and Jim Carden, future backers and board members of HSLDA. At another conference in Sacramento he met lawyer J. Michael Smith, who had also first heard about homeschooling through the Dobson interview and was mulling over the idea of starting a legal firm to represent Christian homeschoolers. At these conferences Moore cleared Farris' lingering doubts about homeschooling (Farris was especially taken with Moore's claim that early schooling damages children). Furthermore, networking with lawyers and potential donors convinced Farris that the time was right to start an organization to defend homeschoolers' rights. In 1983 the Home School Legal Defense Association was born. The problem, however, was that neither Farris nor Smith knew very much about homeschooling legislation or litigation—it was a whole new world to them. That problem was solved when Farris met Chris Klicka.[29]

Chris Klicka's story is a fascinating chapter in the relationship between Rushdoonian ideas and homeschooling. His parents, though not committed Christians themselves, were unhappy with the academic performance of the public schools in Brookfield, Wisconsin, so they sent their son to the local Christian school, which happened to be called Christian Liberty Academy. Though now dissolved, it was in the 1970s a satellite of Lindstrom's school in Illinois, with a self-consciously reconstructionist agenda. Klicka later recalled how "each subject...was taught from a Biblical perspective. The idea was that the scripture is the center and the source of all truth and that you can only gain a proper understanding of science, history, math, politics, and government by looking at it from God's perspective." From there Klicka went to Grove City College just as it was engaged in the early stages of a bitter battle over government regulation. The battle precipitated the Supreme Court's 1984 *Grove City v. Bell* decision, which held that only the admissions office of Grove City had to comply with Title IX regulation. A Democrat-controlled congress responded in 1988 by passing the Civil Rights Restoration Act (known informally as the "Grove City bill") over President Regan's veto, mandating, among other things, that Title IX regulations be applied to all programs of any institution receiving federal aid. Grove City responded by banning all federal aid for students. Such was the political climate at Klicka's *alma mater*.[30]

During the summer of his junior year at Grove City, Klicka left the hothouse environment there for the even more hothouse environment of Gary North's Institute for Christian Economics in Tyler, TX. North, the son-in-law of Rousas Rushdoony, had had a falling out with the patriarch and created his own reconstructionist organization. North has long been a favorite target of liberal critics fearing a Theocratic plot to take down the U.S. government, largely because North is so unapologetically forthright in his aims. "Let us be blunt about it," says North,

> We must use the doctrine of religious liberty to gain independence for Christian schools until we train up a generation of people who know that there is no religious neutrality, no neutral law, no neutral education, and no neutral civil government. Then we will get busy constructing a Bible-based social, political, and religious order which finally denies the religious liberty of the enemies of God.

What would this Bible-based social order look like? For North, it would be a carbon copy of the Old Testament juridical code. "When people curse their parents," for example, "it unquestionably is a capital crime. The integrity of the family must be maintained by the threat of death." Klicka described his time at North's institute as formative.

"I was trained very intensely in the need for us to reform our culture," he said, and his summer in Texas convinced him that he should become a Christian attorney so he could "apply God's principle to this particular arena and protect God's people."[31]

After graduating from Grove City, Klicka attended the fledgling O. W. Coburn School of Law in Tulsa, OK, founded in 1979 by Oral Roberts University (and closed in 1986 when it was folded into Pat Robertson's Regent University School of Law. Klicka was one of a handful of students who were transferred to Regent for their degrees). Klicka notes that while he attended, the school "was battling the monolithic American Bar Association for accreditation so its graduates could become lawyers." One of Klicka's professors there was John Whitehead, who left Coburn to start the Rutherford Institute in 1982. In the summer of 1983 Klicka interned with Rutherford, compiling and analyzing the compulsory education laws of all fifty states so that Whitehead could better represent the homeschoolers he was increasingly being asked to defend. Though Klicka didn't know it at the time, this research project would soon make his career.[32]

In 1985 Klicka became aware of a job opening at Concerned Women for America's legal department, headed by Mike Farris. He applied, but when Farris read his application and learned of his work on compulsory school laws for Rutherford, he called Klicka directly and exclaimed excitedly to him, "You know more about homeschooling laws than I do!" Farris hired Klicka on the spot to become executive director of HSLDA, though Klicka had not yet passed the Bar (it took him three tries). Klicka moved with his wife Tracy to Washington, D.C. and began working.[33]

Unlike the Rutherford Institute, which has always relied on private donations to keep going, HSLDA billed itself from the beginning as membership-based "pre-paid legal defense." Families who joined were charged an annual fee ($100 until 2004 when it was raised to $115) and promised legal representation should they ever need it. For the first few years of its existence, HSLDA plodded along, growing from 200 members its first year to 1200 when Klicka joined the organization. Then enrollment took off. By the end of 1985 HSLDA had 2,000 members. By 1987 it had 3,600. Throughout the 1980s membership figures doubled every thirteen months. By 1994 there were 38,000 members being serviced by thirty-eight full-time employees. By 1996 there were 52,000 members supported by fifty full-time staff. By 1999 HSLDA employed sixty people full time and membership topped 60,000 for the first time. By 2007 HSLDA claimed "eighty plus thousand" member families. What explains such phenomenal growth?[34]

There are several factors that combine to explain HSLDA's rise. First, as we saw in chapter four, conservative Protestants in large numbers were moving in the late seventies and early eighties away from public schools, and these same people were becoming politically engaged. In this sense HSLDA was simply in the right place at the right time. Secondly, many of these newly-engaged Christians had become aware of homeschooling through James Dobson's interviews with Raymond and Dorothy Moore. In those days Moore's endorsement carried tremendous weight, and he endorsed HSLDA. The Moores were at first grateful to have a new organization to lighten their load. Between 1983 and 1986 they referred many parents who were running into trouble with local school officials to HSLDA. Moore gave Farris a list of sympathetic attorneys and served as an expert witness for HSLDA on a number of occasions. Without Moore's early endorsement it is doubtful that HSLDA would have made it.[35]

Thirdly, HSLDA from the beginning hit on a winning fund-raising strategy. Sociologist Matthew Moen has shown how most Christian right organizations founded in the late 1970s and early 1980s did not last, largely because their direct mail/donation funding strategy did not work over the long term. Membership-based organizations, on the other hand, flourished, especially when they were able to hold on to their base while at the same time expanding their vision to include broader issues. HSLDA has been very effective at marketing itself in this way. Using a rhetorical strategy common to other nonprofit organizations trying to keep members committed to the cause, its *Home School Court Report* and other communications have never wavered in producing gripping horror stories of homeschooling families threatened by truant officers, social workers, liberal activist judges, and the National Education Association, along with glowing accounts of the courageous and usually successful defense of these families by HSLDA staff. HSLDA's "Homeschool Heartbeat" radio program debuted in 1991 and within a few years was heard daily on over 900 Christian radio stations. Its lawyers and their wives were frequent keynote speakers and seminar leaders at Christian homeschooling conventions around the country. Klicka, for example, claimed to have spoken at 350 such events by 2006. HSLDA's public relations wing was effective at courting the mainstream press as well. By the 1990s the typical reporter charged with writing the typical homeschooling piece would usually include at least a quotation and sometimes a full family profile provided by HSLDA. As the organization matured, it expanded its scope to include many issues not directly related to homeschooling, a move which irked some members but has proven a successful strategy for keeping the organization relevant long after the fight to legalize homeschooling had been won.³⁶

Finally, and most importantly, HSLDA emerged as the leading homeschooling organization largely because of the reciprocity of an interlocking directorate of homeschool leaders. When Gregg Harris left the Moores, he began to employ Moore's secretary on a freelance basis. She had just begun offering her services as a freelancer, and the aspiring entrepreneur's "first clients were *The Teaching Home*, Gregg Harris and Christian Life Workshops, and the Home School Legal Defense Association. Those were my three customers, and of course at the time they were quite small, they were just starting out." The three organizations shared more than a secretary, however. They shared a vision for cultural transformation via the shepherding of Christian homeschooling families and of course a desire that their startup organizations succeed. Harris plugged Welch's *Teaching Home* at his workshops and urged his listeners to sign up for HSLDA's services whenever he could:

> No matter what the laws may be, apply for membership with the Home School Legal Defense Association. That way, in case you are challenged by local authorities, your legal fees will be covered by the national defense fund.... Annual dues for membership in the Association are a bargain.

Welch, as we have seen, publicized Harris' workshops both in her conference listings and as full-page advertisements in the magazine. She relied on HSLDA for her news and regularly featured Farris as an editorialist. She, too, strongly urged membership in HSLDA and stapled a membership application inside every issue.³⁷

As these three institutions grew together, they came to dominate the homeschooling movement, setting its agenda on a national scale and developing powerful networks to facilitate communication with their thousands of members scattered throughout the country. In 1988 *The Teaching Home* put together the first annual National Christian Home Educators Leadership Conference, open only to leaders of statewide organizations that had an evangelical statement of faith. By 1990 the conference was sponsored as well by the National Center for Home Education (NCHE) as a sort of umbrella group for the closed communion state groups created by HSLDA. By 2002 the conference was strong enough to become an independent entity called the National Alliance of Christian Home Education Leadership. In addition to the annual conference, the National Alliance also has a

Web site where leaders can network, accessible only to approved leaders. "If you are interested in becoming affiliated with the National Alliance," the site says, "you must be a leader of a statewide Christian homeschool organization. Your organization must have written documentation showing that the leadership will be perpetually Christian." The National Alliance maintains a list of approved speakers and their session topics for use by various state conference organizers and is thus the gateway that must be passed through by all who seek access to the thousands of minds and dollars these conferences represent. Sociologist Mitchell Stevens notes that the National Alliance's model has been able to succeed so well because closed communion Protestants are comfortable with hierarchies. For them, "the homeschool world is organized as a pyramid." A small cadre of national leaders gives the talks, writes the books and magazines, provides the vision and sets the agenda, and the state leaders leave the annual conference to disseminate it all to their various regions. The strategy has proven astonishingly effective at providing a powerful public political voice for Christian homeschoolers and at excluding outsiders from sharing in the process. Outsiders, understandably, were not at all pleased with this turn of events. In the concluding portion of this chapter we will look briefly at the way the Christianization of homeschooling was received by those outside the loop and what happened to one insider when she crossed a forbidden line.[38]

The View from Outside

The rapid expansion of homeschooling among conservative Protestants took many veterans from the 1970s by surprise. Many of them grew worried and then angry at what happened to their movement. Two things especially frustrated them: the fracturing of coalitions that had worked together for years across religious and political lines and the emergence of a group of professionals, especially lawyers and curriculum designers, who were changing the face of their "leaderless movement." At first these critics sat by helplessly as their support groups were fragmented, their statewide coalitions marginalized, and their years of activism erased from the historical record in closed communion publications. In the 1990s, however, they tried to fight back.[39]

Who were the outsiders? For the most part they were followers of John Holt. After Holt's death many of his admirers and co-laborers continued *Growing Without Schooling*, most notably editor Susannah Sheffer and publisher Patrick Farenga. In the early 1990s subscriptions continued to rise to a high point of about 5,500, but steady declines thereafter forced its closure in 2001. Another prominent outsider was Pat Montgomery, founder of a free school called Clonlara in Ann Arbor, MI, in 1967. In 1978 she responded to an appeal from Holt and Ed Nagel and began a correspondence program that was used by many unschoolers. By 1993 she had appeared on Donahue, had spent many years "putting out fires" by helping school people see homeschoolers as legitimate, and had won a court case legitimating her approach despite Michigan's law (which at the time required teacher certification for home instruction), all without the assistance of lawyers. Finally, a number of families who by the 1980s were homeschooling veterans converged around Mark and Helen Hegener's publication *Home Education Magazine*, established in 1983. Prominent names here include Larry and Susan Kaseman, longtime homeschooling activists in Wisconsin, and Linda Dobson, a pioneer homeschooler in New York. Other outsiders in this orbit included David and Micki Colfax, famous among homeschoolers for sending three sons to Harvard and writing about it in their 1987 book *Homeschooling for Excellence*, and David Guterson, who was well-known among homeschoolers for his *Family Matters: Why Homeschooling Makes Sense* (1993) well before he penned the bestselling novel *Snow Falling on Cedars*.[40]

One other homeschool leader who by the late 1980s had become an outsider was Raymond Moore. As noted earlier, closed communion leaders, while acknowledging the good work Moore had done for their cause, grew increasingly outspoken in criticism of him as their organizations gained stability. On one hand, some of them looked with incredulity on his Adventism, for the Adventist emphasis on keeping Old Testament dietary laws and other practices seemed to them to compromise the heart of the gospel of salvation by grace through faith in Christ alone apart from works. Historic Adventism, it must be recalled, also saw all other denominations as tainted by the Pagan Antichrist (the Bishop of Rome) and his unbiblical practices, chief among them the mark of the beast itself, Sunday worship. According to many closed communion homeschool leaders, Adventism was a false religion, and Moore was not a true Christian. Secondly, Moore's pedagogy of student-directed activity, delayed academic work, nature study, and service in the community went against the grain of what conservatives had been saying about education for many decades. Many began calling it "unbiblical" and "humanistic," arguing that Moore was "putting research ahead of the Bible." Moore's advice especially annoyed the curriculum providers who stood to make millions of dollars on this market through their phonics and math programs. Richard Fugate, president of Alpha Omega, one of the top Christian curriculum providers, went so far as to publish an extended critique of Moore's pedagogy titled *Will Early Education Ruin Your Child?* While praising Moore's work as a movement activist, Fugate accused him of unwittingly importing the presuppositions of secular humanism into his unbiblical view of childhood nature and learning.[41]

Moore himself did not take this dramatic reversal of his stature lying down. In 1988 he and Dorothy published *Home School Burnout*, a book covering many of the problems homeschooling families often encounter. In the book the Moores provocatively asserted that the biggest enemy to the homeschooling movement was not the public school administrator or the social worker but the coalition of movement insiders who rush children into academic curriculum, hold to radical, un-American antigovernment ideologies, use scare-tactics to drum up membership in their unnecessary associations, and foster division among homeschoolers of good will. He didn't name names, but anyone familiar with the terrain would instantly recognize Gregg Harris and HSLDA as his chief targets. The book was the first of Moore's works not to win general acceptance among homeschool gatekeepers like Cathy Duffy and Mary Pride, and his stature as a movement leader went into precipitous decline. After *Home School Burnout*, the National Home School Convention organized by *The Teaching Home* never again invited Moore or any other speaker who was not part of the closed communion world. By 1994, when his "white paper" that named names and gave sordid details was circulated, Moore had become a pariah among the closed communion leadership. His death on July 13, 2007 was hardly noticed in the homeschooling world. When I spoke with Gregg Harris in August of 2007 it became clear to me that he had not heard the news. After I told him, Harris replied in a soft, sincere tone, "He was a great man, and I believe he's with the Lord."[42]

Several outsiders have tried to create organizations to offset the influence of HSLDA, but they have had a very difficult time doing so, partly because most homeschoolers outside the closed communion world are fiercely protective of their independence and autonomy, partly because the relatively large numbers of Mormon, Catholic, and other non-Protestant religious homeschoolers have their own networks that tend for the most part to keep to themselves, and partly because there simply aren't very many homeschoolers who don't fall into one of the religious categories. Still, they have tried. In 1989, for example,

Connecticut attorney Deborah Stevenson created a local legal defense service for homeschoolers wary of HSLDA's exclusivity and devotion to right-wing political causes. By 2003 her organization had enough momentum to debut nationwide as the National Home Education Legal Defense (NHELD). NHELD's agenda is to ensure that homeschooling remains a state and not a federal affair, to empower networks of local lawyers rather than having paid staff represent various states (as does HSLDA), and to inform local homeschoolers of their legal options so they can take action for themselves rather than becoming dependent on professionals. Members pay twenty dollars a year to subscribe, be they individuals or groups. Since local support groups make up a large portion of its membership, actual figures are only guesses, but Stevenson estimates that by 2007 her organization served perhaps as many as 2,000 people.[43]

The biggest impetus toward creating alternatives to HSLDA came in 1994. In February of that year HSLDA spearheaded an astonishingly successful blitz on Congress to kill an amendment, proposed by Congressman George Miller, to H.R. 6, a house bill reauthorizing funding for the Elementary and Secondary Education Act, which would have required full-time teachers in schools under the jurisdiction of local educational agencies be certified in whatever subjects they teach. HSLDA argued that this amendment could conceivably be interpreted to apply to homeschoolers and initiated a no-holds-barred media alert that produced such a flood of letters and phone calls to Congress that the Capitol switchboard was completely shut down. AT&T estimated that in the eight days leading up to the vote on the Miller Amendment, Congress received between 1 and 1.5 million calls. HSLDA has ever after celebrated this stunning show of force and its resulting defeat of the Miller Amendment and passage of an amendment sponsored by Dick Armey (penned by Mike Farris) explicitly stating that nothing in H.R. 6 applied in any way to homeschooling. But many open communion homeschoolers were not celebrating. They were miffed at HSLDA's unilateral move and claim to be speaking for all homeschoolers, embarrassed at what they took to be a paranoid overreaction to an amendment that in their view had nothing to do with homeschooling, and worried that the Armey Amendment, in introducing "homeschooling" language for the first time into a federal bill, would set a dangerous precedent for the federalization of homeschooling law.[44]

H.R. 6 made many of HSLDA's critics angry enough at their lack of public voice to set aside their differences and create organizations to represent their perspectives. One such organization was the

National Homeschoolers Association (NHA), a coalition of fourteen groups with a total membership of 200. At their first annual conference in 1994 Raymond Moore came and "was there for virtually all of the council meetings," according to an organizer of the event. Moore was trying to recruit the new organization to serve as a political counterweight to HSLDA. "He wants someone to combat Farris and Harris and stand up to them," the organizer noted, but the NHA declined, determining that they were about "diffusing power back to individuals" rather than consolidating it in the hands of a few to be better lobbyists. Moore left disappointed about the prospects of the new group given this organizational strategy, and his premonition proved accurate. After an initial burst of energy, membership declined quickly. The NHA was dissolved in 2000.[45]

The American Homeschool Association (AHA), created by the Hegeners and their publication *Home Education Magazine,* had goals similar to those of the NHA, though in actuality it was little more than an e-mail list and a newsletter. The email list did prove quite popular, and by 1999 many AHA members had learned to trust one another enough to make tentative steps toward coordinated action. For the first time many of these people, previously known to one another only through cyberspace, began to have meetings around the country, and the National Home Education Network (NHEN) was born, committed to a "lateral leadership style" that shunned celebrity or visionary leadership and instead sought to empower all members. While it sounded good in theory, NHEN members did a lot more talking than acting, and even that they did less and less as the years went by. The NHEN discussion board devoted to promoting a public image of homeschoolers' diversity, for example, saw 855 posts in the latter half of 1999 alone. By 2002 there were only 119 for the entire year. In 2003 there were 16. NHEN still exists on the Internet today, but in name only, a sort of virtual ghost town.[46]

What these and other smaller groups did for the most part was grouse about HSLDA. One closed communion insider, who for some time served as executive director of HSLDAs political wing, the National Center for Home Education (and also helped create Bill Gothard's Advanced Training Institute correspondence program), noted with some befuddlement, "there seems to be a tremendous tension generated by those who are in the organizations who are non-Christian, not understanding why the Christians are, you know, not being like Jesus and loving them all." Many other closed communion homeschoolers have wondered at the shrill and sometimes quite vicious attacks pointed in their direction by outsiders. The archives of chat rooms, discussion boards, and e-mail list serves where closed communion types rub virtual shoulders with open communion outsiders offer abundant testimony to this tendency, as do Web sites created to bash HSLDA. Why the rage? First, many outsiders were deeply suspicious of the political aspirations of the religious right in general. They were horrified that their own movement had been co-opted by people they considered to be theocratic Fascists. Secondly, and more importantly, many of the veterans remembered the good old days when a secularist like Holt could share the podium with a reconstructionist like Rushdoony, a Mormon like Joyce Kinmont, an Adventist like Moore, or a Catholic like Phyllis Schlafly, and everyone seemed to get along just fine. The outsiders were no longer invited to speak at the big conventions. Their state groups' fliers were replaced by Christian fliers in *The Teaching Home*. Shut out from the advertising loop, their books and magazines stopped selling like they had in the past. They watched helplessly as people with views they found repugnant made hundreds of thousands of dollars off their movement, enjoyed the media spotlight, and gained the ears of high-level political figures. Their antipathy was mixed with not a little envy, making for a particularly bitter stew. What made it all the more difficult to choke down was the fact that their criticisms, though sometimes overstated and almost always overheated, often had some legitimacy.[47]

To illustrate, we will close this chapter with the biggest scandal to have emerged out of this feud between closed and open communion homeschoolers. That this story is so little known outside of the small world of open communion outsiders is testimony to just how marginal they had become by the mid 1990s. Cheryl Lindsey was a Washington State native who met her first husband, a member of the Seattle Black Panther Party, while she was a student at the University of Washington. She was drawn to his intelligence and charm but reaped a marriage marked by physical brutality. After a few years and two children she divorced this man, who eventually was sentenced to life in prison on several counts of assault and battery. She married another African American man named Claude, who was also abusive. In the 1970s the couple turned to fundamentalist Christianity, settling eventually in the Calvary Chapel movement (the same group that had evangelized Gregg Harris), a loose affiliation of churches growing out of the West-coast hippie counterculture. Calvary Chapel churches are deeply biblicist and practice a form of church governance that gives considerable autonomy to local pastors, many of whom exert great authority over their members' personal lives. The Lindseys attended such a church, and over the years Cheryl, influenced by this community, Bill Gothard, and others, grew to repudiate her earlier dabblings in feminism and to embrace natalist ideals associated with what she called "home centered living:" homespun clothing, home birth, courtship, house church, canning and bread making, full-quiver family size, and so on. Lindsey began homeschooling in 1983, and in 1989 as a mother of nine began publishing her own magazine, called *Gentle Spirit*, with an initial circulation of seventeen. *Gentle Spirit* emphasized homeschooling as an entire lifestyle, with articles on godly femininity, submission to male authority, and lots of practical advice on rural living and child-rearing. After a Dobson appearance and plug, her magazine and speaking career on the homeschooling circuit took off. By 1994 *Gentle Spirit* had a circulation of 17,000 and Lindsey was keynoting some of the biggest conventions in the country. But by 1995 her career as a Christian homeschool leader was over.[48]

The circumstances surrounding her fall are complicated. Lindsey's husband had abused her for many years, and they were frequently separated even as the children multiplied and Cheryl Lindsey deepened her commitment to and involvement in home-centered living. By the spring of 1994, after Claude had relocated to New Orleans, Cheryl filed for divorce and met another man. "I had been publishing articles encouraging Christian women to be chaste, obedient, submissive, keepers of the home; now I seemed to be turning against everything I had been standing for," Cheryl later wrote. Nevertheless, she had outstanding speaking commitments to keep, so in the summer of 1994 she keynoted the Christian Home Educators of Ohio (CHEO) convention. By then word had gotten around to some homeschooling leaders about her personal problems, and there were whispers that her new lover Rick Seelhoff might have even accompanied Cheryl to the conference. Gregg Harris, Sue Welch, and CHEO executive director Michael Boutout all aggressively sought out information about Lindsey's personal life, and her pastor, Joe Williams, provided it. To confirm that Rick Seelhoff had indeed accompanied Lindsey to the conference, Gregg Harris called Seelhoff, and, while taking care not to lie outright, led Seelhoff to believe he worked for the hotel in Columbus where Lindsey had stayed during the conference. Harris asked if Seelhoff might have lost his credit card at the hotel, and when Seelhoff answered, "yes," Harris quickly closed the conversation and hung up.[49]

With the adultery confirmed, many closed communion leaders conspired to drive Lindsey out of the business. Sue Welch created a packet of materials including a letter of discipline from pastor Williams and other materials verifying Lindsey's divorce. She sent this packet out to every state leader affiliated with her national network, and all of Cheryl's speaking engagements were cancelled. Boutout and Pastor Williams compiled a list of "proofs of repentance" that they mandated for Cheryl should she wish to be restored to fellowship. These included requirements that she refrain from public speaking, give up her beeper and P.O. Box, stop answering her phone, turn over her business and bank account to a third party, agree not to consult lawyers, and stop publishing *Gentle Spirit*, filling outstanding subscriptions with *The Teaching Home*, Welch's publication. Welch also sent her packet to Mary Pride, whose newsletter *HELP For Growing Families*, founded in the same year as *Gentle Spirit*, was on the verge of being discontinued because *Gentle Spirit* dominated the market Pride was targeting. When Pride became aware of the issue, she saw an opportunity and began aggressively courting *Gentle Spirit* subscribers through Internet discussion boards and advertisers through direct communication.[50]

For the next two years Lindsey was harassed any time she tried to restart her business. Finally, in May of 1997 she sued everyone she could for violating the Sherman Antitrust Act: The Williamses, Calvary Chapel, Sue Welch and *The Teaching Home*, Gregg Harris, CHEO and Michael Boutout, and the Prides. Suddenly, the harassment stopped. Boutout, Harris, Calvary Chapel, and Mary Pride all settled with Lindsey out of court for amounts all parties agreed to keep confidential. Only Welch would not, and *Seelhoff v. Welch* went to trial in 1998 in a U.S. District Court in Tacoma, Washington. In September, a unanimous jury found that Welch had entered into conspiracy to restrain trade and ordered a payment of $445,000 in damages. Since it was an antitrust case, the award was tripled. Welch was forced to pay over $1.3 million plus lawyer fees to Cheryl Lindsey Seelhoff.[51]

The fallout of this case makes for a fitting close to this chapter on the rise of closed communion homeschooling leadership. Among open communion publications and organizations the case was huge. *Home Education Magazine* devoted many pages to extensive coverage and analysis of the case. The Hegeners, Linda Dobson, and others who were associated with their work painstakingly transcribed and web-published the entire trial record and were quick to point out the many instances where Christian leaders were caught hedging the truth during their depositions by evidence that surfaced over the course of the trial. For all their efforts, however, Cheryl Lindsey and the *Gentle Spirit* controversy are today almost completely unknown in the closed communion world. Longtime veterans may remember the issue, but it has left no permanent mark. The defendants paid the money quietly and have refused to discuss the matter since then. Chris Klicka, for example, when asked by a reporter to comment on the Lindsey affair, replied, "Oh, that adulterer. I haven't read her stuff, so I can't respond to it." Discussion of the case was forbidden by moderators on most Internet boards devoted to closed communion homeschooling. None of the closed communion publications reported on the trial or its outcome. Nevertheless, though the closed communion world continued on as if nothing had happened, subtle changes did take place. The most obvious casualty was *The Teaching Home*. Welch was never able to recover from the verdict's devastating financial blow, and *The Teaching Home* ceased publication in 2002. Though Welch remained involved in national homeschool networking, she kept a low profile after the ruling. Less difficult to document was a quiet shift in momentum away from hard-core closed communion conservatism in the movement as a whole. Many of the most extreme natalist publications saw declines in their subscription base after 1998, causing many of them to go under in the ensuing years. Gregg Harris stopped giving his annual seminars shortly after the ill-fated Columbus conference, and by the time of the jury verdict he had shifted his amazing entrepreneurial energies from homeschooling to local church growth, leadership training, and newsprint evangelism. HSLDA, while its membership numbers continued to grow, shifted its emphasis in many respects as well, as we will discuss in the last chapter of this book. Many analysts and commentators have noticed that in the years since 1998, homeschooling as a movement has shifted toward the American mainstream: conferences are no longer so dominated by fundamentalist curricula; women in denim jumpers and headcoverings are now joined by women in slacks and tank tops; A-list celebrities, extreme athletes, and minorities who homeschool now receive more media attention than conservative Christians. While it is impossible to say just how much of this shuffling and mainstreaming is the result of the *Gentle Spirit* decision, it would not be too much of a stretch to see its public rebuke of closed communion chutzpah as a reality-check to movement insiders, who learned that the parallel institutional universe they had spent the past two decades creating was still subject to U.S., not biblical, law.[52]

Chapter Seven

Making It Legal

One of the greatest achievements of the homeschooling movement was the legalization of homeschooling in the 1980s and early 1990s in every state in the country. Yet this very important story has seldom been told outside the annals of homeschoolers' own publications. It is a difficult story to tell, for two reasons. First, since U.S. education law is predominantly a state affair and not a federal one, there are actually fifty stories to tell. These fifty stories interface in complicated ways as well: court cases in one state are cited in others, legislative trends become contagious, and national organizations often exert significant influence on local politics. Secondly, the sharp division between closed and open communion homeschoolers we chronicled in the last chapter has left a strong imprint on the way various homeschooling groups themselves have described what happened. If one reads the closed communion memoirs and artifacts, one might learn that "the modern homeschool movement was started through a miraculous moving of the Holy Spirit" that began around 1983, prior to which time homeschooling was legal "in only five states," or perhaps was banned "in all but three states." It "was treated as a crime in almost every state" and parents who homeschooled "frequently faced jail terms and the loss of their children to foster care." But "because of HSLDA, which has won virtually every legal battle it has fought, and because of the warm support of Republican legislators, home schooling is now legal in all 50 states." If you read HSLDA's critics, on the other hand, you get the impression that, while there were some problems, on the whole homeschooling has always been fairly easy to do in most places so long as homeschoolers were civil, and in those places where it was not easy, most of the heavy lifting had been done before HSLDA even came on the scene. Furthermore, critics allege that most of HSLDA's work since then has actually made matters worse for homeschoolers. A faithful account of the legislative and legal history of homeschooling must therefore accomplish the difficult task of reducing the various state stories to some sort of manageable order while avoiding the pitfalls of partisan polemics. That is the goal of this chapter.[1]

Happily, the surge of evangelicals into homeschooling in the early 1980s got many legal scholars, reporters, and educational researchers interested in this new and controversial phenomenon. Chris Klicka was not the only lawyer poring over antiquated state compulsory school laws to try and sort out what was legal. The following account is derived largely from the body of work produced by these scholars, much of which was published in obscure law journals in the 1980s and 1990s. When reading through the literature, one does not get a sense that most of these people had strong opinions one way or the other about the value of homeschooling—they were just doing their best to figure out what the state statutes and court cases had to say on the matter and to chart trends they saw emerging.

There are two basic issues to cover here. The first is constitutional. Is homeschooling a right guaranteed by the U.S. constitution? Many homeschoolers have long argued that it is, usually citing either the First Amendment right to the free exercise of religion or the Fourteenth Amendment due process clause and the "right to privacy" that has developed from it. The second is statutory. Is homeschooling permissible according to the various state statutes concerned with the education of children? In this chapter we will look first at the constitutional question and then at the much more complicated statutory question.[2]

The Constitutional Battle

In the late 1970s, when John Holt first began organizing the motley band of people teaching their children at home, many discussions were held about the best strategy to pursue to make homeschooling clearly legal. Holt and many others learned to go to their local libraries, read the archaic school laws, and think up creative interpretations that would allow for homeschooling. Some of Holt's readers wondered if it might be more efficient and effective to bypass this messy statutory terrain and the dense thickets of conflicting case law and simply try to secure a Supreme Court decision that would interpret the constitution as providing a clear sanction for homeschooling. Holt himself did not favor this strategy, but it has been tried on many occasions. To date, however, the Supreme Court has not agreed to hear any case specifically pertaining to homeschooling. That has not, however, stopped lawyers from making constitutional arguments in lower courts. HSLDA lawyers have been especially forceful in this regard, and their writings repeatedly assert that "homeschooling is not a privilege granted by the state. Homeschooling is a right that is guaranteed by a higher law: the Constitution of the United States through the First and Fourteenth Amendments."[3]

The Fourteenth Amendment argument may be the stronger of the two. The Supreme Court has long recognized parental rights as part of the constitutional right to privacy. *Meyer v. Nebraska* (1923) and *Pierce v. Society of Sisters* (1925), affirmed a constitutional right "to marry, establish a home, and bring up children." *Prince v. Massachusetts* (1943) stated, "it is cardinal with us that the custody, care, and nurture of the child reside first in the parents, whose primary function and freedom include preparation for obligations the state can neither supply nor hinder." *Roe v. Wade* and *United States v. Orito* (both in 1973) found there to be "fundamental" privacy rights in the domains of marriage, procreation, motherhood, child-rearing, and education in the Constitution. In these and many other cases the Supreme Court has consistently resisted efforts by government to infringe upon the due process rights of parents. On the other hand, the Supreme Court has also consistently upheld the power of states to "compel attendance at some school" (*Meyer v. Nebraska*) and to make sure private schools, in the words of *Board v. Allen* (1968), provide "minimum hours of instruction, employ teachers of specified training and cover prescribed subjects of instruction." In homeschooling these two judicial principles collide, leaving many lower courts at a loss over how to proceed. The precise line between what a state can and cannot regulate in terms of a child's education has never been established by the Supreme Court. Lower courts have typically not found a Fourteenth Amendment right to homeschool; not in Illinois (*Scoma v. Chicago Board of Education*, 1974), New York (*In re Franz*, 1977), Michigan (*Hanson v. Cushman*, 1980), New Mexico (*State v. Edgington*, 1983), Arkansas (*Murphy v. State*, 1988), or Maine (*Blount v. Department of Educational and Cultural Services*, 1988). The one case that did, *Perchemlides v. Frizzle* discussed in chapter five, though it became celebrated for this very reason in the homeschooling community, was in fact an unreported lower court decision with no precedent-setting power. Its judicial impact, according to two legal scholars, "has been markedly limited." While many activists have argued that the Fourteenth Amendment, in the words of Alma Henderson, "protects a parent's decision to teach a child at home," to date very few courts have agreed.[4]

The First Amendment argument has also been fronted on many occasions. The claim is that compulsory schooling infringes on religious free exercise when a family's religion forbids government involvement in children's education. *Wisconsin v. Yoder* (1972), the famous case wherein the Supreme Court upheld the right of Amish families to cease formal education after the eighth grade, is often cited as a precedent for this claim. But the Court was very careful to circumscribe its decision in *Yoder*, holding that "probably few other religious groups or sects" could qualify for a similar exemption. Still, homeschoolers have often employed the First Amendment defense, and usually they have lost, as they did in Illinois, (*Scoma v. Chicago Board of Education*, 1974), West Virginia, (*State v. Riddle*, 1981), Alabama (*Jernigan v. State*, 1981), North Carolina (*Duro v. District Attorney*, 1983), Texas (*Howell v. State*, 1986), Ohio (*State v. Schmidt*, 1987), and North Dakota (*State v. Patzer*, 1986 and *State v. Melin*, 1988). But in one state the free exercise defense has worked twice. The first time, *State v. Nobel* (1980), a Michigan district court found no compelling state interest to infringe on the free exercise rights of the Nobel family, who claimed that their fundamentalist Christian religion forbade them from sending their children to schools certified by the state of Michigan. The fact that it was a lower court case and that Mrs. Nobel was herself a certified teacher, despite her religiously based rejection of state certification, made the case unique and non–precedent setting. But in a second decision, *People v. DeJonge* (1993), the Michigan state Supreme Court became the first in the nation to accept the free exercise defense. The case is remarkable in several respects. By 1993 Michigan was one of the last holdouts to require homeschool teachers to possess a state teaching certificate. After repeated failures at all lower court levels, HSLDA finally secured a 4 to 3 ruling that allowed the DeJonge family to educate their children at home without certification as a constitutional right grounded on the First Amendment right to free exercise of religion. The same day, also by a 4 to 3 margin, the same court rejected the Fourteenth Amendment argument in *Michigan v. Bennett*. The dissenting opinion in *Bennett* reads a lot like the majority opinion in *DeJonge*, and vice versa. The difference between the two outcomes was the swing vote of Justice Charles Levin. According to Chris Klicka, who was a defense attorney for the case, a reporter for the *Detroit Free Press* learned from Chief Justice Cavanagh that Levin had planned on voting against the DeJonge family, but at the last minute he changed his mind. "Why?" asked the Chief Justice. "I don't know. I just want to change my vote," responded Levin. As a result, writes Klicka, "the printers had to scramble and change the majority opinion to the dissenting opinion and the dissenting opinion to the majority opinion." This decision was HSLDA's greatest courtroom achievement. Michael Farris called the decision "the most significant victory—not just for

HSLDA, but for homeschoolers ever, anywhere, anytime, anyplace." And given the 11th hour conversion, it is not surprising that Klicka would see in it the hand of God: "We know what happened-even if Justice Levin did not. The Bible tells us in Proverbs 21:1, 'The king's heart is like channels of water in the hand of the LORD; He turns it wherever he wishes.' That is what God did. He turned Justice Levin's heart."[5]

Homeschooling advocates often cite *Perchemlides* and *DeJonge* to claim a First and Fourteenth Amendment right to homeschool. But the truth is that these two cases are outliers. "The clear weight of authority" rejects such arguments. Legal analyst Perry Zirkel noted in 1997 that there has long been "a general constitutional trend...that disfavors homeschoolers' claims based on First Amendment free exercise and Fourteenth Amendment parental liberty." Zirkel, like Holt in the 1970s, counseled homeschoolers to focus their efforts not on ill-fated constitutional arguments but on local issues, especially on legal opinions offering favorable interpretations of state compulsory education statutes and on legislative efforts to rewrite those statutes when they did not favor homeschooling. "The future of home instruction," wrote legal scholar Henry Richmond III in 1980, "rests largely in the hand of the state legislatures." Subsequent events have proven him right.[6]

The Statutory Battles

By 1918 every state in the union had a compulsory school law. Most of these were written around the turn of the century with the aim of getting children, especially immigrant children, off the streets, out of the factories and mines, and into schools where they could be taught the English language and American values. As we saw in chapter three, legal challenges to these laws in the early and mid-twentieth century were rare, and those that were mounted yielded no clear consensus either for or against home education. Until the late 1970s when homeschooling quickly morphed from being a rare and isolated experience to a fairly common one, state legislatures had not paid much attention to their aging compulsory education statutes. The new homeschoolers, looking for wiggle-room, did.[7]

What they found surprised them. State laws, while nearly identical in many respects, dealt with domestic education in different ways. At the dawn of the movement, fourteen state statutes said nothing at all about education at home but usually mentioned the acceptability of children being taught in a private school. Fifteen explicitly mentioned home instruction in one way or another. The remaining twenty-one contained phrases like "equivalent instruction elsewhere" or "instruction by a private tutor" that could be read to imply recognition of home education as a legitimate option. The thirty-six states with either explicit or implied provisions for home instruction differed markedly over the specificity of their rules governing non–public school instruction and over establishing who was in charge of it all. Some were very vague. Some empowered local school boards to govern such matters. Some statutes established robust requirements. Six even required that any teacher of children, regardless of venue, be certified by the same standards the state used to certify public school teachers.[8]

As homeschoolers learned all of this, they began to develop different strategies in different states. In states with laws that did not mention home instruction at all, homeschoolers had two basic options. Some of them tried to argue that their home schools were private schools, for every law had something to say about private schools, and the Supreme Court had long upheld the rights of parents to choose them. Others signed up for a correspondence program connected to a recognized private school. Both approaches could bring trouble. Most of the pre-1978 case law had concluded that home schools were not private schools, often on the grounds that they did not properly socialize children. In the words of *State v. Hoyt* (1929) in New Hampshire, home-tutored children missed out on "association with all classes of society." Similar decisions were rendered in California (*People v. Turner*, 1953), again in New Hampshire (*In Re Davis*, 1974), and Kansas (*State v. Lowry*, 1963 and *In Re Sawyer*, 1983). On the other hand, some earlier cases had rejected the socialization argument and found that home schools could be seen as legitimate private schools. Such was the case in Illinois (*People v. Levisen*, 1950 and *Scoma v. Chicago Board of Education*, 1974) and New Jersey (*State v. Massa* 1967). In the latter case the court argued that "to hold that the statute requires equivalent social contact and development would emasculate this alternative and allow only group education, thereby eliminating private tutoring for home education." By the 1980s rulings tended in the general direction of finding that homeschools do count as private schools, and that they should only be evaluated by academic, not social, standards. This has been the result in Georgia (*Roemhild v. State*, 1983), North Carolina (*Delconte v. State*, 1988), Texas (*Leeper v. Arlington Independent School District*, 1987), and Colorado (*People in Interest of D. B.*, 1988).[9]

Having established that home schools were private schools, or having signed up via correspondence with an actual private school, homeschoolers still faced the same challenges familiar to other private schools. In earlier chapters we have noted the dramatic rise in independent Protestant education in the 1970s and 1980s. This growth led to many conflicts and not a little litigation over how tightly local school districts and state legislatures could regulate academic matters in the new schools. Here again constitutional arguments, usually of the First Amendment variety, were fronted, with mixed results. In many cases courts upheld state authority to require certified teachers or other "reasonable" requirements as not placing an undue burden on free exercise, but in some cases state regulations were deemed "so pervasive and all-encompassing that total compliance with each and every standard by a nonpublic school would effectively eradicate the distinction between public and nonpublic education," as *State v. Whisner* (1976), an important Ohio decision, put it. The *Whisner* decision is interesting in that, though the verdict was unanimous, judges were divided over whether constitutional principles were at stake or whether the case should have been decided on purely procedural grounds. Again, absent a clear word from the Supreme Court, the constitutional issues involved in such cases remain hazy and open to judicial caprice.[10]

In any event, homeschoolers taking the private school option were thrust into this maelstrom, dependent on the good will of local school officials. Local officials by the mid-1980s typically did not harbor good will toward homeschoolers, as several surveys of principles, teachers, and school superintendents have documented. Some, like the National Education Association's Robert McClure, believed that "it's important for children to move outside their families and learn how to function with strangers," fearing that home instruction would undermine commitment to American pluralism. Others worried that lax pedagogy would not equip children for the world of college and work where, as the Virginia State Board of Education noted in 1982, "discipline and control is more objective." Many raised the classic socialization argument, noting, in the words of Omar P. Norton of the Maine Department of Education, "instruction in isolation cannot compare with a child being educated in a group." But by far the greatest concern was the threat parent-teachers represented to claims of professional expertise. John Cole, president of the Texas Federation of Teachers, fretted that "if anyone can teach, teaching will, indeed, no longer be a profession." Donald Bemis, a supervisor of public instruction in Michigan, memorably expressed the view of most of his peers, "If you need a license to cut hair you should have one to mold a kid's mind."[11]

In states with either explicit or implied home education statutory provisions, homeschoolers faced a legal climate little different than that of states without such provisions. As Holt noted, "What most of these state laws boil down to is that you can educate your child at home only if the local school board says you can—and most say no." Many state laws stipulated how many hours a day and days a year a student must be taught. Most had some sort of formal approval process in place, usually at the discretion of local school boards or superintendents. Most empowered local districts to set regulations for curriculum content. And, as noted, six required state teacher certification. Given the immense variability across states and even between districts within states, homeschoolers' strategies varied considerably. Many of the unschoolers of the 1970s tried to keep a low profile. Some didn't register their children at all. Others successfully built bridges with local school people. Often a parent's attitude could make all the difference. Elaine Mahoney of Cape Cod, MA, for example, was permitted to homeschool in 1978 largely because of her demeanor. William Geick, the assistant superintendent overseeing her case, noted, "Mrs. Mahoney came to me not as a parent angry at the school system, but as a parent with a different philosophical approach, based not only on her opinion but on sound recommendation." When Geick and Mahoney presented her home education plan to the five-member school committee they were impressed. One member recalled, "I wasn't very receptive to the plan until I met Elaine. She impressed me as a serious, conscientious woman who was able to give this time to her children." Mahoney later praised the school committee who signed off on her plan, "I respect them because they care. Because of that, anything is possible."[12]

Mahoney's was not an isolated case. Just before the massive influx of conservative Christians into the movement, Raymond and Dorothy Moore wrote, "We have found that in about 80 to 90 percent of all cases where parents have kept their children out of school until the ages of eight or ten or later, local public school administrators and primary teachers...are understanding." Pat Montgomery reported similar numbers, "Of the thousands of families Clonlara has served, relatively few—twenty-eight to be exact—have ever had contact from local officials that could not be handled by a simple phone call or letter." Of the twenty-eight, only eighteen had to go to court. Seventeen won their cases, sixteen without a lawyer. "The record shows that families who home educate, by and large, have little to fear from officials." Ed Nagel, coordinator of the National Association for the Legal Support of Alternative Schools, said much the same thing in 1982, "Today, about 95 percent of all home-study court cases are being decided in favor of the parents."[13]

But upon the entrance of thousands of angry Protestants who were convinced that public schools were "Satanic hothouses," relations with officials were strained considerably. HSLDA's Scott Somerville noted that while many unschoolers successfully cooperated with school officials, religious conservatives were often aggressively antagonistic, leading to showdowns. "Some rejected public education as 'godless' and thought of school officials as secular humanists who were bent on godless mind control." When such people confronted school officials with home education plans, "the legal battles began in earnest." Such soured relations affected unschoolers as well. Pat Montgomery, who just a few years before had noted how cordial relations between her clients and local school districts were, recalled the mid-1980s as the "'look over your shoulder' time," when school officials were "more often than not, hostile; their urge to regulate was percolating."[14]

Several types of legal battles ensued. In states that implied the acceptance of home instruction by mentioning "other equivalent instruction" than public or private schools, the question again was raised as to whether equivalency extended to social interactions or ought to be limited to academic matters. New Jersey illustrates well the general national trend. In early decisions (*Stephens v. Bongart*, 1937 and *Knox v. O'Brien*, 1950) courts interpreted equivalency to mean social experiences and thus disallowed home instruction. But in *State v. Massa* (1967) the New Jersey court reversed itself and required only academic equivalency. Yet limiting equivalency to academic matters still meant that state courts had to confront the extent to which government could regulate curriculum and teacher preparation. Here the most effective legal strategy homeschoolers found to counteract regulations they found onerous was the "vagueness" defense, the argument that state statutes are unconstitutionally vague, thus contributing to the randomness of policy in various districts. This vagueness defense has succeeded in many states, including Wisconsin, (*State v. Popanz*, 1983), Georgia, (*Roemhild v. State*, 1983), Minnesota (*State v. Newstrom*, 1985), Missouri (*Ellis*, 1985), Iowa (*Fellowship Baptist Church*, 1987), and Pennsylvania (*Jeffrey v. O'Donnell*, 1988). In such instances state legislatures had to revise their compulsory school laws, and the results usually made homeschooling easier. But the vagueness defense failed in other states. Alabama (*Jernigan v. State*, 1981), Iowa (*State v. Moorhead*, 1981), and West Virginia (*State v. Riddle*, 1981) all upheld such clauses when they were tested.[15]

Finally, in states with explicit clauses regulating home instruction, many kinds of legal battles ensued. First, many homeschoolers again argued that such statutory requirements, especially state certification, infringed on their First or Fourteenth Amendment rights. In nearly every instance such arguments have failed. Secondly, some homeschoolers in states with specific language about home instruction have nevertheless tried to have their home school classified as a private school instead, because private schools in many states faced less restrictive regulation. In nearly every instance this strategy failed as well. Others argued that their home schools were doing just fine and didn't need to adhere formally to state regulation. This strategy likewise failed, even when courts agreed that the parents were doing a good job. Finally, many cases have been concerned with whether or not various homeschooling arrangements meet the statutory requirements the law sets out, garnering results as varied as the nature of the cases. In short, given the failure of constitutional challenges to state statutes, courts have generally held parents to the letter of the law, to the great frustration of homeschool-friendly lawyers. Chris Klicka "learned quickly that the victory was not always in the courtroom. In fact, it usually wasn't in the courtroom. I learned, if at all possible, to avoid the courtroom because it is completely unpredictable." On the other hand, courts have usually held school districts to the letter of the law as well when their zeal to prosecute outpaced their legal authority. This has been true especially in the higher courts, which frequently overturned convictions sustained by lower courts friendly with school district personnel but not as conversant with legal precedent and statutory interpretation.[16]

Faced with repeated failure in the courts both in their constitutional arguments and in their creative strategies for sidestepping statutory regulations, homeschoolers in many states decided that if they could not work with the laws as they existed, they would have to change them. As Holt put it in 1983, "we must now greatly increase our efforts...to get good home schooling laws passed in the legislatures." Thirty-seven states created or updated home education language in their compulsory school laws in the years between 1981 and 1991, many of them in response to pressure from well-organized and vocal homeschooling advocates energized by court cases that upheld extant law, some of them in response to courts finding statutes unconstitutionally vague. Each state's story is a unique combination of local personalities and conditions mixed with national trends and organizations, and of longstanding regional culture mixed with anomalous, unpredictable events. Perhaps the best way to understand what occurred nationally is to look briefly at several examples of what happened locally.[17]

Winning in the Legislatures

Before discussing statutory reform it should be noted that in many states such issues simply never came up. In California, for example, though there is no explicit statutory language permitting home education and though the higher courts have historically rejected the claim that home schools are private schools, thousands of homeschoolers have registered as private schools every year for decades without much trouble. In 1986 an unreported district case (*People v. Darrah and Black*, 1986) found California's law unconstitutionally vague, but the case was not appealed and nothing was done about it. By the late 1980s the California Department of Education (CDE) was regularly advising "parents who wish to educate their children without local school district involvement" to "file an affidavit for exemption as a private school." While in more recent years the CDE has voiced concerns about this policy, any attempt to change it now would reap the whirlwind. In short, though California has more homeschoolers than any other state, it has no law, nor does it seem to need one. Several other states have similarly been able to accommodate home instruction without legislating it. In fact there has been not a little controversy among homeschoolers in many states over efforts by some (often led by HSLDA) to push for new laws. While some homeschoolers feel the need for an official sanction for their behavior, others, especially unschoolers, tend to prefer the vagaries of unregulated practice. New legislation usually grants legitimacy but increases regulation.[18]

In most states, however, there are now explicit statutes governing home instruction. Many of these came as a result of court cases that made it clear that such provisions were necessary. Georgia's history provides a clear example of how courtroom drama often begat drama in state legislatures. In Georgia, three cases in the late 1970s and early 1980s created momentum for reform. The first, *Blankenship v. Georgia* (1979), pitted a single mother and her two boys against the educational establishment in a made-for-media David and Goliath story. Patty Blankenship had been using the Accelerated Christian Education curriculum for three months when she was met at her door by a police officer, probation official, and the local public school principal who threatened her with imprisonment, loss of custody, and a fine of $100 per diem if she did not send her children to school immediately. She fled with the two boys, but was arrested, jailed, and released on a $1000 bond. When two officers came the following Monday to take her boys away, they found the house vacated. Blankenship secured the services of a young lawyer named Ted Price who worked *pro bono*. Raymond Moore testified on her behalf as Price argued that her school qualified as a legal private school under Georgia's law, which made no mention at all of home education. The trial ended in a hung jury, but the school district let her go for fear that a retrial might result in a ruling that the compulsory school law was unconstitutionally vague.[19]

Sixteen months after the Blankeship trial, Terry and Vicki Roemhild removed their three children from the Georgia public school system because they found it less academically sound than the schools they had patronized in North Carolina and because their Worldwide Church of God beliefs caused them to object to the school's holiday celebrations and immoral student culture. They had read Holt's and Moore's books and sent letters of intent to local and state officials as advised, but, getting no response, they simply started homeschooling. Nineteen days later, school officials got the county District Attorney to file misdemeanor charges. The Roemhilds' state-appointed attorney James Irvin chose to use the vagueness defense rather than a First Amendment challenge, but the judge ruled against the family because he personally felt that the parents, both of whom lacked college degrees, were unqualified to teach their children. The Roemhilds appealed.

Meanwhile, the state Board of Education, worried over the increasingly commonplace claim made by Georgia homeschoolers that their efforts qualified as private education, set about constructing a narrow definition of a private school. To qualify, said the Board, a school must have at least fifteen students enrolled for 180 days in a building used primarily for educational purposes with at least one teacher possessing an accredited college degree. This proposal was scheduled to be voted on by the Georgia state legislature in February of 1983. When word got out about the proposal in January, several groups quickly formed to lobby against it. Connie Shaw, a Mormon who subscribed to *Growing Without Schooling* quickly organized "Georgians for Freedom in Education" with seventy charter members. E. Lewis and Patricia Gibson, both of whom had attended Shaw's meeting but who were wary of working with Mormons and others, organized a closed communion "Christians Concerned for Education." Finally, Steve and Ann Nichols, disciples of Holt and Moore, organized a local chapter of the National Council of Parent-Educators. All three groups lobbied hard for the defeat of the Board's new definition, aided by the even stronger voice of Georgia's private school interests. Many private schools feared the new regulations, especially the requirement that teachers must come from accredited colleges, since many of them employed graduates from unaccredited Bible schools. The efforts of all of these groups were successful in derailing the board's proposal.[20]

Even as the state Legislature was arguing over the School Board's proposed definition, a third case was being tried. Wimbric and Marion Padgett were Holiness Pentecostals whose children had attended a small segregation academy until its implosion in 1981. Since that time they had been homeschooling their two children using the ACE curriculum. They were prosecuted by the Telfair County School Board and secured the services of attorney J. Dan Pelletier who worked for them *pro bono*. At trial, Raymond Moore and Rousas Rushdoony testified on their behalf free of charge, their transportation and accommodation being paid for by the homeschooling groups previously mentioned. The presence of such prominent figures in a sleepy Georgia courtroom made for great news and impressed superior court judge Phillip West, who ruled that the Padgetts were indeed operating a private school.[21]

As this third case was winding down, the Roemhild appeal was also underway. To support the Roemhilds and draw attention to homeschooling in Georgia, Raymond Moore and Phyllis Schlafly hosted a two-day conference in Atlanta in October of 1983 that drew a crowd of 500, including some of the state's most influential government and media people. Only days after the conference, on October 25, the Georgia Supreme Court, in a 4 to 3 decision, overturned the Roemhild conviction. The court ignored First Amendment arguments but declared the state's compulsory education law unconstitutionally vague, noting especially a similar Wisconsin Supreme Court ruling handed down earlier in the year (*State v. Popanz*, 1983). The vagueness of the law violated the due process rights of the defendants, the Court said, because, "a criminal statute must be sufficiently definite to give a person of ordinary intelligence fair notice of the behavior which is required or prohibited."[22]

With the law struck down for vagueness, momentum shifted again to the legislature. The months after the *Roemhild* decision saw much discussion of homeschooling. On January 10, 1984 state

Senator John Foster, chair of the Senate Education Committee, proposed a new compulsory education law that explicitly excluded home schools from the definition of private schooling. Governor Joe Frank Harris and other legislators convinced him to change it, and successive drafts got looser and looser in regulating home education, due largely to unrelenting pressure from homeschoolers on the Governor and key legislators. After eight drafts, a bill was finally brought before the Senate. It required home educators to submit a letter of intent, to teach only their own children for at least four and a half hours a day and 180 days a year, to submit standardized test scores every three years, monthly attendance records, and basic curriculum to the School Board. Finally, the bill required homeschooling parents to pay a fine of $100 for breaking any of these rules. In the Senate an amendment was added requiring home educators to possess either a high school diploma or a GED. The state's four black Senators, led by Julian Bond, at first wondered if the bill was a smokescreen for resisting desegregation, but they became convinced during debate that the movement derived primarily from religious and academic motives and thus supported it. The final vote was 52 to 1 in the Senate, and, despite heated objections from state Superintendent of Schools Charles McDaniel, it passed in the House as well (146 to 24). On April 3, 1984 Governor Harris signed it into law. In the first year of the new law, 653 families filed letters of intent for 1,071 children, a number that increased steadily every year thereafter. By 1993 there were 6,137 Georgia families submitting letters for 10,523 children—8 percent of all Georgia students. By 2006 there were 38,531 children formally recognized as homeschoolers in Georgia.[23]

Between 1982 and 1988, twenty-eight states passed new homeschooling legislation. Many of these laws have stories that bear striking resemblance to Georgia's, so much so that it would be tedious to recount them all. But here follows three more stories to give a sense of the range of similarities and differences to be found. In Oregon, homeschoolers were tired of the arbitrary treatment to which they were subjected by a law that required homeschooling be approved by local school districts. Dennis Tuuri, an outspoken reconstructionist pastor, founded the Parents Education Association Political Action Committee (PEAPAC) in 1983 and was able to secure passage in 1985 of the Home School Freedom Bill, which legalized home education so long as school districts are notified and children are tested in grades 3, 5, 8, and 10 by "a qualified neutral person." Oregon public school officials, taken off guard by this surprise legislative victory, mounted an effort to repeal the bill in 1987. PEAPAC sprung to action, taking advantage of the thick connections that the Oregon Christian Home Education Association Network (OCEAN) had been building among closed communion homeschoolers in Oregon since Gregg Harris' first seminar there in 1985. OCEAN sent the word out, *The Teaching Home* lent its voice to the cause, and Oregon homeschoolers overwhelmed the state legislature with more phone calls and letters than the legislators had ever seen on any other issue. During hearings on House Bill 3342, the 140-person room was always filled to capacity and the halls outside lined with homeschoolers voicing their objections to this effort to undo their legislative gains. In contrast, the two groups who supported the bill, the Oregon School Board Association and the Confederation of Oregon School Administrators, sent one professional lobbyist each. The bill never made it out of committee. Several subsequent attempts by the Oregon Department of Education to toughen homeschooling regulations were likewise thwarted by the overwhelming force of activist homeschoolers.[24]

Like in Oregon, Maryland homeschoolers turned to legislative relief out of frustration over harassment from the Maryland Board of Education, but unlike in Oregon and many other states, their efforts did not result in a new law. The Maryland statute had always required public school attendance unless children were "otherwise receiving regular and thorough instruction in the courses of study normally taught the children of the same age in public schools." By 1980 enough Marylanders were providing such instruction in their homes that Manfred Smith, a public school teacher and disciple of John Holt, was able to organize the Maryland Home Education Association (MHEA). In 1984 the children of Kathleen Miller, who had been homeschooling with the Calvert curriculum for a year, were prosecuted by the Ann Arundel County Public Schools for truancy. MHEA secured attorney Ray Fidler for the defense. Fidler argued that the Miller children, both of whom were testing well above grade level, were in fact receiving "regular and thorough instruction." The circuit court agreed. After the loss, the Board of Education and local school districts grew even more hostile, so MHEA and several other interested individuals got Ann Arundel delegate John Gray to introduce legislation in the State House of Representatives that would legalize homeschooling with no restrictions whatsoever. The bill passed the House with modifications but failed in the Senate where the Board of Education, now in panic mode, promised Senators that they would work with homeschoolers to draft new regulations acceptable to everyone and thus that a new law was not necessary.

Over the next year the Board of Education worked with Manfred Smith, staff from the Calvert School, and many other interested parties to carve out options for homeschooling in Maryland. An independent option was crafted requiring an annual intent to homeschool be submitted to the local Superintendent of Schools, who was also authorized to review a portfolio of "relevant materials" up to three times a year if desired. Maryland homeschoolers were also permitted to enroll in correspondence programs that complied with state regulations.[25]

Just to the North, Pennsylvania homeschoolers were facing similar issues. In 1982 Jim and Gloria Gustafson were worried about their second grade son Jonathan who was academically advanced but socially fragile. They heard the Moores on Dobson, read *Home Grown Kids*, and started homeschooling their kindergartener David. Later they added Jonathan as well. When word got out that a professor at Messiah College in central Pennsylvania was homeschooling, other like-minded families began contacting him with questions and a support group was born, later named HAHA (Harrisburg Area Homeschoolers Association). It quickly became clear to Gustafson and others that there was no consensus among Pennsylvania's 501 school districts over how to deal with homeschoolers. The extant law allowed for schooling by "a properly qualified private tutor" if the district superintendent approved. But what did it mean to be properly qualified? Some superintendents were friendly and let families do their thing. Others were overtly hostile. The Gustafsons themselves were strongly discouraged from homeschooling by their local superintendent, but he grudgingly allowed them to do it. Other families were less successful. Some even moved to other districts known to be more friendly to their cause.[26]

At the same time as the Gustafsons and other like-minded families were beginning to find one another in Central Pennsylvania and trade information about the vicissitudes of superintendent opinions in their region, others were doing the same in other parts of the state. Homeschoolers in Western PA got organized after a visit from John Holt in March of 1982 to a television station in Pittsburgh. Many homeschoolers came to the station, met one another, and began a mailing list. Howard and Susan Richman brought thirty-four families out to their farm in June and began a newsletter to keep everyone in touch. From its first issue, sent out to fewer than 100 addresses, the newsletter grew to become a leading statewide magazine. By 1995 *PA Homeschoolers* was being mailed to 2,000 families. Its early issues were very largely preoccupied with the random nature of homeschool policies in various districts and what homeschoolers could do about the situation.[27]

Meanwhile, over on the eastern side of the state, Ann Cameron ran into trouble trying to homeschool her daughter LaAnna. After writing a "kindly letter" to her local superintendent she received a response that threatened fines and imprisonment if she pulled her child. She scheduled an appointment with the superintendent, who "railed upon" her for half an hour. After she described for him the bad publicity he would get when the media learned of his persecution of a young mother who was doing a fine job educating her daughter at no cost to the taxpayers, he relented. But the experience convinced Cameron that something had to be done to curb the caprice of local school leaders. She began a support group that started with ten families but quickly grew to several hundred.[28]

As the movement grew, so did the suspicions of superintendents, so much so that many homeschooling leaders began murmuring among themselves about possible strategies for relief. Cameron contacted every regional support group she knew, inviting them all to a statewide meeting to address the growing conflict between homeschoolers and local districts. Howard Richman was one of those who responded to Cameron's summons. Like Cameron, he too had come to the conclusion that something had to be done, but he had a different solution than she. At this meeting the first signs of what would eventually become a serious internecine conflict between Pennsylvania homeschoolers emerged. Richman brought to the meeting a draft of a proposed homeschooling bill that might be brought before the Pennsylvania legislature. Cameron was "aghast" at his proposal, full as it was with regulations for parents. She and some other leaders felt that government had no business whatsoever being involved in family life and that homeschoolers should push to have the compulsory school law declared unconstitutional. Richman argued that an outright assault on compulsory education would not work. His faction won, and he and Tom Eldredge spearheaded an effort to lobby state legislators for the passage of a homeschooling bill.[29]

The lobbying effort took far longer than anyone anticipated, partly due to the volatile political climate in Pennsylvania during the mid- and late-1980s and partly because of strong, organized, and vocal opposition by the Pennsylvania Department of Education. Legislative battles spilled over to local district policies as more and more local superintendents got tough on homeschoolers. Districts that had been friendly suddenly started prosecuting families for truancy amidst a climate of mutual distrust and disdain. By the beginning of the 1988 school year a new law still had not been passed. HSLDA reported that sixty-four of its Pennsylvania members had been challenged by their districts that fall, causing them to dub Pennsylvania the "Worst State of the Year" in terms of homeschooling freedoms. HSLDA had earlier filed a case on behalf of several Pennsylvania families, and in August of 1988 a federal court declared the tutoring provision of Pennsylvania's school law unconstitutionally vague (*Jeffery v. O'Donnell*), though it rejected HSLDA's First and Fourteenth amendment arguments. The decision gave new urgency to the bill homeschoolers had been lobbying for in Harrisburg for the last four years. Finally, after unrelenting pressure from homeschoolers at the state capitol, including breakfasts and meetings featuring Raymond Moore, Michael Farris, and other homeschooling luminaries, both the State Legislature and Senate unanimously passed what became Act 169.[30]

The law, full as it was of compromises, is one of the more restrictive in the nation, requiring homeschooling families to file annual affidavits with their local district, maintain a portfolio including an annual evaluation by a sanctioned outsider, and submit standardized test scores in grades 3, 5, and 8. While many homeschoolers who spent hours and hours lobbying for the bill were understandably euphoric at its passage, others worried that the law represented "surrender to The State" by extending "a godless jurisdiction across the threshold of the door of the home." Many moms who had been homeschooling for years felt like the new policies infringed on their freedom, as every field trip now had to be fitted into curricular mandates and documented in the portfolio. Sharon Lerew, for example, always "felt like something was hanging over my head." Others made the best of the situation, believing that peace of mind gained by the new uniform policies outweighed the cumbersome restrictions.[31]

As in other states, as soon as the law was passed, homeschooling coalitions quickly fragmented. The Richmans, probably the most conspicuous Pennsylvania leaders, were Jewish, and they not infrequently collided with "several homeschoolers from Lancaster County" whose Christian exclusivism and unwillingness to compromise with government they found distasteful. The Lancaster County Home Education Association (L'CHEA), formed in 1983, was a closed communion group but worked with the Richmans and others in the early years to secure passage of the homeschool bill. As the restrictions mounted, however, L'CHEA grew increasingly frustrated with Howard Richman's capitulations to state officials. Over time, L'CHEA grew into a statewide organization and changed its name to Christian Homeschool Association of Pennsylvania (CHAP). CHAP has always been closely allied to HSLDA, Gregg Harris, and the National Alliance of Christian Home Education Leadership that grew out of Sue Welch's National Conferences. CHAP's convention and curriculum fair, begun in 1986, has grown exponentially into one of the largest in the country, regularly attracting over 8,000 people who browse mostly Christian materials and attend a full slate of sessions, the vast majority offered by Christian leaders. In recent years tensions have mounted as CHAP has tried to loosen homeschooling requirements by lobbying for legislation with which the Richmans disagree. HSLDA has also filed lawsuits on behalf of Christian homeschoolers in Pennsylvania prosecuted for truancy after refusing to submit the required affidavits. The suits claim that the extant law violates Pennsylvania's Religious Freedom Protection Act as well as the First and Fourteenth Amendments. To date such initiatives have failed, but tensions remain high among many Pennsylvania homeschoolers.[32]

These are only a few of many states that revisited their compulsory school laws in the mid 1980s. In nearly every case the end result was a more explicit law that gave homeschoolers legal status but also regulated them to a greater or lesser degree. Sometimes the leading players were Christian conservatives, sometimes not. If there is a general national trend, it is that legal and legislative battles tended to be sponsored more by open communion coalitions in the early 1980s, gradually giving way to closed communion leadership by the end of the decade. The passage of the Pennsylvania law in late 1988 was one of the last initiatives backed by a wide range of groups spearheaded by leaders who were not conservative Protestants. By 1989 most states had come to terms with homeschooling in one way or another and a general peace settled on the land. But there were still a few holdouts even into the 1990s, and in these states it was the closed communion leadership, and especially HSLDA, that did much of the heavy lifting. To conclude the chapter we will look briefly at some of these final battles.

The Last Holdouts

Texas had historically been very friendly toward domestic education, with a long history of private tutoring on the frontier. In the early years of the twentieth century, when Texas' compulsory school law was passed, at least 70 percent of Texas children were being taught in their homes either by their own parents or by tutors, often circuit riders. The 1915 law required children to attend either public or "private or parochial school." In 1981, however, the Texas Education Agency crafted a new policy stating that "Educating a child at home is not the same as private school instruction, and therefore, not an acceptable substitute." Around 150 homeschooling families around the state were prosecuted, 80 going to trial. By 1985 homeschoolers had had enough. Several Texas homeschooling families led by spitfire Southern Baptist lawyer Shelby Sharpe filed a class-action suit against all 1,050 school districts. The Texas Education Association and State Board of Education tried to get the case dropped by crafting new guidelines for private schools, but this only enervated the powerful private school sector. At public hearings on the proposed guidelines in April of 1986, over 6,000 citizens massed in Austin to protest government intrusion into private education, an event that has gone down in Texas lore as "the Austin TEA (Texas Education Association) Party." The TEA's regulations were summarily rejected by the state legislature.[33]

In January of 1987 *Leeper v. Arlington* went to trial. Rousas Rushdoony, Raymond Moore, and reconstructionist author Samuel Blumenfeld all testified in the homeschoolers' behalf. In April, Justice Charles J. Murray sidestepped constitutional free exercise and due process arguments and concluded that homeschools are indeed private schools according to the Texas statute. The TEA appealed. In November of 1991 the Texas Court of Appeals upheld the lower court's ruling without changes. The TEA appealed again. In June of 1994 *Texas Education Agency v. Leeper* was finally put to rest by the Texas Supreme Court, which unanimously upheld the lower court's ruling on statutory grounds, again ignoring First and Fourteenth Amendment arguments. Homeschoolers in Texas were thus brought under Texas' very loose private school provisions, making it one of the least regulated states in the country.[34]

There are several points to be made about *Leeper*. First, this long battle proved devastating to the Texas Education Agency, whose recalcitrance ultimately cost the school districts of Texas over $700,000 in plaintiff legal fees, not to mention the cost of their own lawyers. This lesson was not lost on other states. Secondly, it left lasting battle scars between public school leaders and Texas homeschoolers. A 1999 doctoral dissertation found deep distrust still among Texas superintendents and principals, most of whom were extremely dubious of the quality of education imparted by homeschooling, and also among homeschoolers, most of whom were not open at all to any sort of cooperation with public schools, beyond the free use of school libraries. Finally, the case reflects the tension between closed communion and open communion homeschoolers over the legacy of their movement. When the case was filed HSLDA was still in its infancy, and throughout it played only a very minor role in the proceedings. But that has not stopped HSLDA lawyers from repeatedly declaring *Leeper* as "HSLDA's first big win" and referencing it in membership drives. Many of HSLDA's critics see this as a classic example of HSLDA's frequent attempts to take credit for work done by others. Larry Kaseman, an outspoken critic of HSLDA, called attorney Sharpe to find out about HSLDA's role and was told "the extent of HSLDA's involvement in the case was when Sharpe asked an HSLDA attorney to take the stand as a witness merely to identify HSLDA as an organization."[35]

HSLDA's role may have been minimal in Texas, but it was very significant in most of the other states facing serious legal challenges in the 1990s. In South Carolina, for example, it was a key player. State law there allowed for home instruction that was "substantially equivalent" to public or private schooling but left it up to local districts to determine what this meant. Some required a high school diploma for parents, others required teacher certification. Some simply rejected all requests. In 1983 a family court judge overturned one district's rejection of a homeschooling application (*Calhoun County Dept of Ed v. Scott Page and Susan Page*). This action angered school officials across the state, who felt their authority was being usurped, and drew attention to what many of them saw as a growing threat. The next year, Zan Tyler, frustrated that her local principal would not allow her to delay her son's kindergarten enrollment, applied to her school board for permission to homeschool. She was denied permission. But the district was messing with the wrong woman. Tyler's father, one of the most influential businessmen in the state, contacted U.S. Senator Strom Thurmond, who personally flew to Columbia to advocate for his friend's daughter, and Tyler's rejection was immediately overturned by the State Board of Education. This bald show of force enraged many school officials. In 1985 a task force was organized to draw up new statewide guidelines for homeschooling, including requirements that all homeschooling parents must have a four-year degree, use only state-approved texts, and take all standardized tests given in public schools. Tyler sprang to action, alerting the 400 families she had on her mailing list. Their letters forced public hearings on the proposed guidelines. More than 350 parents and other advocates converged on Columbia to offer four hours of testimony, but the State Department of Education sent the recommendations unchanged to the General Assembly. More hearings were held, and this time over 700 homeschoolers showed up. The Department of Education's proposals were rejected. Homeschooling-friendly legislators then introduced a home education bill that was finally passed, but with an amendment requiring parents without a Bachelor's degree to pass a test called the Education Entrance Exam (EEE) before being permitted to homeschool.[36]

HSLDA filed a class-action lawsuit on behalf of many members who believed the EEE was not a valid test of homeschooling competence. They lost in the lower court in 1989 but appealed. Meanwhile, Tyler tried an end-run around the EEE requirement by founding the South Carolina Association of Independent Home Schools (SCAIHS), an umbrella organization that sought to offer its own accreditation to home-study programs and thus bypass the district approval process. The state Attorney General ruled that SCAIHS was illegal, and HSLDA filed another lawsuit. In December of 1991 the South Carolina Supreme Court overturned the lower court's decision and declared the EEE an invalid test for homeschoolers. Working with this momentum, SCAIHS successfully shepherded a bill through the state House and Senate, thanks in large part to the behind-the-scenes advocacy of Tyler's well-connected father and continuous pressure from grassroots homeschoolers. By 1992 South Carolina homeschoolers could legally have their homeschooling programs approved by Tyler's SCAIHS organization as an alternative to school district oversight.[37]

Yet here again tensions between closed communion and open communion homeschoolers appeared. Zan Tyler had long been a close ally of HSLDA, frequently writing articles for their *Home School Court Report* and speaking at closed communion conferences and conventions. SCAIHS, while not officially a Christian organization, nevertheless had a board made up entirely of Evangelical Christians and an overwhelmingly Christian membership, leading to "an association with a Christian culture," in the words of founding member Jim Carper. For the first several years of its existence SCAIHS required families using its accreditation service to be members of HSLDA. All of this was very frustrating to some South Carolina homeschoolers who, for whatever reason, did not wish to flee school district oversight only to be subjected to the oversight of SCAIHS (and especially to be forced to join HSLDA). In the mid-1990s several outsiders worked to establish alternatives to SCAIHS accreditation, and in June of 1996 what has become known as the "third option" law passed the South Carolina legislature, granting groups other than SCAIHS the same authority to accredit homeschooling programs. Over the next decade, third option providers proliferated at a remarkable rate. Most of them offered their services for a fraction of the cost SCAIHS charged, and relations between these new groups and SCAIHS have not always been cordial. Third option organizations often saw SCAIHS as a monopolistic behemoth while SCAIHS worried that many third option groups' lax accountability requirements gave South Carolina homeschooling a bad reputation. Nevertheless, though SCAIHS was not an enthusiastic endorser of the third option law and though its leadership has often suffered stinging criticism from several third option leaders, influential figures like Tyler and the Carpers have consistently defended the right of third option groups to exist before skeptical state officials. In more recent years several explicitly Christian third option organizations have emerged, several with ties to HSLDA, thus muddying the waters even further.[38]

HSLDA also played a central role in two final battleground states of the early 1990s. In 1991, after years of protests, rallies, court cases, task forces, and failed bills, many of them spearheaded by HSLDA's Mike Farris, Michael Smith, and Chris Klicka, Iowa finally passed a homeschooling law that overturned a longstanding policy requiring teacher certification of all homeschooling parents. And last of all came Michigan, where, as we have noted, HSLDA won its most significant victory in 1993 when the State Supreme Court accepted its First Amendment constitutional argument by a one-vote majority. This decision, along with two others handed down in 1993, established homeschooling as a legitimate form of private education in Michigan. It also set the stage for Michigan's 1996 homeschool law, which, by eliminating "all notice and reporting requirements," made Michigan one of the most permissive homeschooling states in the country. But here again some homeschoolers balked at HSLDA's tactics. Many Michigan activists, including Clonlara's Pat Montgomery, believed that the 1993 court victories had secured enough freedom and did not want to risk introducing formal home school language into state statutes that might lead to further regulations down the road. But HSLDA and the closed communion Michigan group INCH (Information Network for Christian Homes) successfully guided a "stealth" bill through the legislature without the knowledge of the well-organized open communion homeschoolers in that state. Montgomery and others tried to repeal the law in 1997, but they failed.[39]

Despite such internal squabbles, by the middle of the 1990s homeschooling had become legal and popular in every state of the union. As Mary McConnell put it, "in most jurisdictions, and with varying degrees of grace, the educational authorities have surrendered to the homeschoolers." Why did homeschoolers win? Many homeschoolers will tell you that they won because, in Chris Klicka's words, "God was on their side." But homeschoolers had other allies as well. In the first place, as Mitchell Stevens has noted, homeschooling activists enjoyed a "favorable institutional ecology" due to the decentralized nature of U.S. educational policy. Homeschoolers could build beachheads in friendly states and expand from there. Secondly, as Stephen Bates has argued, homeschoolers possessed "deep-seated passion" stemming often from a religious zeal and always from parental love that their opposition simply could not match. They were able to mobilize that energy through their vast networks of support groups, mailing lists, periodicals, conferences, and national organizations. Thirdly, as James Cibulka has observed, bureaucracies tend to prefer accommodation to confrontation. School leaders tended to back off when homeschoolers and their lawyers challenged them. Homeschoolers quickly discovered that school officials' "most effective weapon is bluffing," and when homeschoolers refused to be pushed around and threatened legal action, they usually got their way.[40]

Critical to their success is the fact that homeschoolers won in the court of public opinion. A 1985 Gallup poll found that 70 percent of Americans thought homeschooling should not be legal. A decade later, however, Gallup found that 70 percent of Americans believed homeschooling to be a valid educational alternative. There are two reasons for the shift. First, homeschooling families have proven to be very effective at public relations. Their children's success and the arguments they have amassed against predictable questions have won over the average American. When asked if homeschooling properly socializes children, for example, homeschoolers have learned that most Americans readily assent when it is noted that public schools do not exactly excel in this domain themselves. Many Americans are also impressed by the manners, facility of speech, and maturity of so many homeschooled children when contrasted to the typical mall-rat saturated in peer culture. In general it can be asserted that homeschoolers have been very adept at making arguments that resonate with many deeply held American beliefs: "our belief that all people are individuals, with rights; our suspicion that 'experts' are not as trustworthy as common sense, and our worries that government is too intrusive and does not serve us very well."[41]

A final reason for the public acceptance of homeschooling is media coverage. Back in 1983 John Holt noted that "the press and other media have been virtually without exception friendly to home schooling and home schoolers; I cannot recall a single interview or report that was hostile." This trend has continued unabated. Michael Apple, an outspoken critic of homeschooling, noted in 2000 that media outlets usually present homeschooling as "a savior, a truly compelling alternative to the public school system that is presented as a failure." Isabel Lyman's examination of 340 articles on homeschooling in print media published between 1985 and 1997 found that the overwhelming majority of articles depicted homeschooling "in a positive light," concluding that such stories have "played a major role in publicizing this countercultural trend to mainstream America." Reporters, like Americans more generally, love rooting for the underdog, and the story line of persecuted parents taking on the educational establishment and winning has proven hard to resist. The following anecdote from Rebekah Pearl, daughter of home school pioneers Michael and Debi Pearl of No Greater Joy Ministries, illustrates well the role media coverage has played in homeschooler victories over educational administrators. The setting is rural Tennessee in 1982:

> Social Services had gotten wind of our home education, and we were given a court summons. The judge and a few power-hungry and small-minded individuals assured Dad that his children would be taken away from him and put into state care. Dad came home, and within half an hour had three television stations and three newspapers scheduled to do a story on us. They came out to our rather fine home in the Shelby forest and filmed me (eight-year-old Rebekah) playing the piano, my brother Gabriel working in the shop with Dad, and my four-year-old brother Nathan swinging on a rope over the pond. They talked about Dad's Bachelor of Science education, his artwork (he was a professional landscape painter) and showed clips of our schoolroom with posters and desks all tidy and organized. (Actually, we did most of our school in the yard or the kitchen—but it sure looked good.) They put forward the question, why didn't the state just test us, and leave us alone if we tested up to state standards? Dad's strategy more than succeeded. The state let go of us in a panic, and families all over the place started calling us for information about homeschooling.[42]

Since the mid-1990s, homeschooling has grown, diversified, and gained even more legal ground. Occasionally a state representative or Board of Education will suggest tightening control of home education, but vigilant homeschool watchdogs quickly whip up a vocal reaction and such proposals are hastily defeated. To cite but one example, in 2002, Connecticut State Representative Cameron Staples sponsored a bill that would have required homeschoolers to possess at least a high school diploma and have their curriculum approved by their local superintendent. E-mail alerts were quickly sent out by Diane Connors of the Connecticut Homeschool Network, and over 1,000 people showed up at public hearings to voice opposition to the bill. Only one superintendent showed up to support it. The legislature hastily killed the bill. On the public relations front, repeated exposure to homeschooling via constant and overwhelmingly positive media attention has made the practice seem less strange and exotic to the average American. The final chapter of this book will examine several recent trends that have turned homeschooling into a major, and increasingly mainstream, force in American education, so much so that we may be dealing no longer with a self-conscious protest movement but something more like the routine domestic education of the past.[43]

Chapter Eight

Homeschooling and the Return of Domestic Education, 1998–2008

"I never really told anybody about my music at school, only my really close friends," Cheyenne Kimball told *People Magazine* in 2006. "Then [school officials] actually aired the show around the whole entire school, and that caused a lot of problems. I was a straight-A student and all of a sudden I didn't want to go to school anymore because of the things people were saying. That's why I'm homeschooled now." Cheyenne, winner of NBC's *America's Most Talented Kid* at age twelve, recording artist, and star of her own MTV show, is just one of many high-profile Americans whose educational choices make it clear that homeschooling has gone mainstream. Bethany Hamilton, a homeschooled surfer girl, became an instant celebrity when she survived a shark attack in 2003 that cost her an arm. Her tragedy and inspiring recovery have been leveraged into a "miniempire" of merchandise, including her own perfume, accessory line, fiction, advice books, and autobiography. Maeghan Kearney and her brother Michael became famous as homeschooled prodigies. "I was so bored with regular school," said Maeghan, "I would finish all the assignments on Monday and have the rest of the week off. So my mom home-schooled me and I finished the fifth grade to high school in two years." Maeghan graduated from college at age fourteen. Her brother Michael holds the Guinness World Record for being the youngest person ever to complete high school, college, and graduate school.[1]

Movie stars Will and Jada Pinkett Smith, married in 1997, homeschool their two children along with Will's nephew. Why? "For flexibility," Jada told an *Essence* reporter, "so they can stay with us when we travel, and also because the school system in this country—public and private—is designed for the industrial age. We're in a technological age. We don't want our kids to memorize. We want them to learn." R & B sensation John Legend was homeschooled by his mother, who also worked as a seamstress, until high school, allowing him to skip two grades and immerse himself in gospel music. Her sacrifices paid big dividends when her son's first album, *Get Lifted*, won three Grammy awards.[2]

Homeschoolers have been tearing up the Scripps Howard National Spelling Bee and the National Geographic Bee for well over a decade. Since 1991, the first year the Scripps organization began keeping track, 10 to 15 percent of its national participants have been homeschooled. A homeschooler first won the competition in 1997. In 2000 the top three finishers were all homeschooled, and the winner, George Thampy, also placed second in the Geography Bee. George's two brothers and sister have all placed highly in various bees as well, making the family something of a legend in the world of competitive spelling and geography. When asked to account for his family's success, George explained, "most schools have a system of learning that is calculated…to suck the very marrow out of a subject. Too often, the quest for grades impedes the quest for learning, while in homeschooling, without grades, kids are free to learn unhindered." In 2007 homeschooler Caitlin Snaring became the first girl in seventeen years to win the Geography Bee, a feat she credited to homeschool flexibility. "I integrated all my subjects with geography," she said, studying sixty hours a week to prepare: "I wanted a girl to win this." Homeschoolers won the National Geographic Bee five times between 1998 and 2007.[3]

One could go on for many pages listing the achievements of homeschooled children and adding to the list of notables choosing this route. Homeschooling has come a long way since the late 1970s when Phil Donahue could ignite impassioned debate just by suggesting the idea. As many of the above examples illustrate, homeschooling is no longer the sole preserve of the leftist counterculture or Christian fundamentalism. While the Thampys and many other well-known homeschoolers are committed Christians, they are famous not for their religious beliefs but for their worldly success. If there is a theme uniting the disparate trends, personalities, and movements discussed in this final chapter it is this steady move of homeschooling from the fringes to the mainstream of American life. In the introduction, I made a distinction between "homeschooling" as a compound word and separated phrases such as "home education" and "domestic education," explaining that for me "homeschooling" designates the effort to teach children in the home as a deliberate alternative to and rejection of institutional schooling. By the late 1970s that effort had become a viable political movement. By the dawn of the twenty-first century, that movement had won its legal fight and gained acceptance by most Americans. Increasing popularity, however, has diluted much of the radical spirit that enervated the early adopters of the 1970s and 1980s, leading to a sense of profound ambiguity among many movement veterans over what has become of their creation. In this chapter we examine recent demographic and commercial trends in "homeschooling" that are leading us back to "home schooling," to the use of the home to educate not as a gesture of protest but simply because it makes the most sense for some families given their situations. We consider the impact of these changes on the homeschooling movement's self-image and conclude with some reflections on the meaning of homeschooling, and home schooling, for American society.

Demographics and Marketing

Home-based education's increasing popularity coincides with the continuation and acceleration of trends described in chapter four that laid the groundwork for its emergence in the first place. It is most popular among populations residing in what one scholar called "moderate growth zones of internal migration," which is to say, in suburbia, especially the Sunbelt region. The great majority of the 1.5 million homes built in the United States every year are single-family suburban dwellings. Family size may have shrunk, but home size has grown from an average of 800 square feet on a 5,000 square foot lot in the 1950s to 2,300 square feet on a 13,000 square foot lot today. Bigger houses separated further and further from commercial districts and other public buildings have led to a dramatic increase in the number of cars owned and miles driven since the 1980s, as well as a growing affinity for stay-at-home work and "e-tailing," or Internet-based shopping. In 2000, for example, 19.6 million people worked in telecommunications jobs, many of them stay-at-home moms living on the rural fringes of suburban expansion. The clear national trend toward privatized living that has been with us since the end of World War II has continued unabated. Home education fits naturally into this model.[4]

Reliable nationwide numbers continue to be difficult to obtain, but recent publications from the National Center for Educational Statistics have provided the best estimates to date of the number of children being taught at home. Based upon samples from the massive National Household Surveys Program, the NCES estimates that in 1999 there were around 850,000 homeschoolers, a figure that increased to roughly 1.1 million by 2003, a 29 percent jump in four years. Movement leaders continue to suggest even higher estimates.

Brian Ray, director of the HSLDA-affiliated National Home Education Research Institute, claimed somewhere between 1.9 and 2.4 million homeschoolers in 2006. A less speculative sense of the movement's growth can be seen from figures kept by some states. Virginia, for example, had 3,816 registered homeschoolers in 1990. By 2005 the number had grown to 23,252. Maryland saw a similar growth from 2,296 in 1990 to 20,676 in 2002. In these and other states annual growth rates have begun to slow somewhat after over a decade of dynamic expansion. There are even hints in some states that homeschooling rates may have peaked and are even declining a bit. In Pennsylvania, for example, there were 24,415 reported homeschoolers in 2003, the largest figure the state had ever seen. But in 2004 the number of registered homeschoolers dropped for the first time ever, to 24,076. In 2005 it dropped again to 23,287, a decrease of 3.3 percent from the previous year. The most likely explanation for the declines here and in some other states showing similar drops is the increased use of home-based public charter schools, often called "cybercharters" because of their extensive use of online curricula. Many families taking advantage of this educational choice had previously been homeschooling independently.[5]

Recent research has also revealed a considerably more heterogeneous population of homeschoolers than earlier and more limited studies had found. Polling data has found far higher rates of minority homeschooling (20 percent in one study) than previously believed. The U.S. Department of Education estimated that in 1999 almost two thirds of homeschooling families had three or more children, 64 percent of homeschooling households made $50,000 or less, and a majority of homeschooling parents had not completed college. Several recent studies of parental motivation have found academic concerns to outweigh religious reasons in the choice to homeschool. Guillermo Montes' analysis of the National Household Survey data found that 70 percent of respondents cited a nonreligious reason as the top motivator in their decision to homeschool. Homeschoolers whose motivations are primarily religious have certainly not gone away (41.31 percent in Montes' study), but they are now joined by many who do so for all sorts of reasons, ranging from concerns about special education, to bad experiences with teachers or school bullies, to a proliferation of time-consuming outside activities, to worries over peanut allergies. Many such parents, unfamiliar with the niche culture of conservative Protestantism, have often felt unwelcome in support groups created in the 1980s and 1990s by Christians, leading to an explosion in recent years of groups without any doctrinal platform.[6]

Several in-depth qualitative studies of homeschooling families have revealed interesting details about homeschool life. A longitudinal study of seventeen families in central Ohio found a consistent pattern of pedagogical development. In the first year of homeschooling, mothers often relied on a prefabricated curriculum and sought to do everything by the book. By the second year they had become more flexible ("eclectic" being the term of choice) and tended to engage their children in more outside activities. By the third year of homeschooling, parents had often become more like facilitators and the kids were largely in control of their learning. Another study found, contrary to popular perception, that the average homeschooled child spends only two years outside of the classroom before returning to some sort of traditional school. Another found that homeschooling rates tend to be highest when mothers with high educational attainment and aspirations for their children live in school districts with significant percentages of low-income families. Finally and most interestingly, given the history we have recounted painting such a sharp distinction between closed communion and open communion homeschoolers, several studies are finding that "no simple division exists between religiously-motivated and academically-motivated parents." Longtime movement observer Cathy Duffy some years ago faulted homeschooling research for overly polarizing the movement "while missing the middle ground where I find most home schoolers." Scholarship is beginning to catch up with her intuition. In a recent review of the literature, Ed Collom concluded, "studies of parental motivations indicate that homeschooling has become more mainstream and that there are a host of middle-grounders with varying rationales."[7]

As the movement has grown and matured, so have its institutions. Homeschooling is now big business. By 2001 over 500 conventions were being held every year around the country, seventy-five of them attracting over 3,000 people each. The leading periodicals *Homeschooling Today*, *Practical Homeschooling*, and *Home Education* had a combined circulation of over 350,000 a month. Newsletters, online message boards, and blogs have proliferated madly. A few companies have successfully harnessed this roiling market and ridden it to great profits. In 2001 Paul and Gena Suarez, building on the momentum generated by their thriving eBay business selling used homeschooling curricula, launched *Old Schoolhouse Magazine*, a quarterly 200+ page glossy whose production standards rival those of any mainstream checkout-counter mag. Like many of them, *Old Schoolhouse* is mostly ads—pages and pages of full-color advertisements for a dizzying array of homeschooling products. Scattered amidst the ads are feature articles, often interviews, which tend to read like advertising copy themselves. *Old Schoolhouse*, while clearly a Christian publication, has quite a different vibe to it than the older generation of Christian publications like *The Teaching Home* or *Practical Homeschooling*. It regularly features curricula and organizations that are not explicitly Christian and runs articles about "unschooling" and other progressive-oriented pedagogies, all in an effort to be the magazine of choice "for homeschooling families everywhere." Over 30,000 such families were receiving the magazine by 2007.[8]

Even more successful has been the "Home School Headquarters" initiative of Appalachian Bible Company, launched in 1999. With the bewildering proliferation of curricula and specialty items made available to homeschoolers by the late 1990s, many Christian booksellers were just as confused as parents were as to what products they should carry. Appalachian began a program whereby local Christian bookstores could purchase one copy of Appalachian's entire stock of products (an initial investment of about $6,000 in 2004) and pay an annual fee of around $700 to establish a "home school headquarters" in their store where families could peruse resources before buying them. Whatever a family chose to purchase was then ordered directly from Appalachian, thereby saving local store owners the task of researching the latest products, guessing what would sell, and being stuck with unsold inventory. Participating stores also receive referrals through Appalachian's mail-order service and many other perks. Appalachian began with sixty-five stores. By 2004 they had signed up 250 stores for the service. Customers like being able to peruse actual products at the bookstore before buying, and owners love the business the deal brings in and the headaches it avoids.[9]

Established Christian presses have watched all of this growth with great interest, and some have tried to colonize it. In 2001 Broadman and Holman, a Nashville-based Christian publishing house with Southern Baptist roots, bought out Loyal Publishing and a few other "mom and pop" homeschooling curricula and hired veteran South Carolina homeschool leader Zan Tyler to help them gain market share. "The growth of home-schooling will continue between 7% and 15% annually," said Tyler in 2004. "Right now, it's worth about $850 million in curriculum and resources." To date, however, Broadman and Holman's efforts have not borne fruit. In 2006 the homeschooling division was put on hiatus, staff relocated to other divisions, and Tyler released. Other established presses have had similar difficulties. Veteran publishing house Thomas Nelson has tried to cash in, but found homeschooling, in the words of their vice president, Dee Ann Grand, "a vapor market because we couldn't grab it." Zondervan's senior marketing director Kathy Needham noted in 2004 that "home-school is now a category we're intentional about," and the venerable publisher has tried to market some of its Bibles and historical fiction to homeschoolers. Multnomah, Crossway, and Bethany House have made similar efforts. But to date none of them has had much success. Homeschoolers typically prefer to purchase products on the recommendation of other homeschoolers, and a product's reputation has a lot to do with the life story of its creator, which homeschoolers want to know. The market has proven largely immune to traditional strategies like the slick and somewhat mawkish mailers and catalogues established presses have long used to sell their products. Homeschoolers much prefer direct online purchasing through such outlets as Christian Book Distributors, Discount Home School Supplies, Rainbow Resource Center, Timberdoodle, and dozens of other sites, and they tend toward frugality. Scores of online sites exist to facilitate used curriculum sales, and homeschoolers have become legendary among librarians for their patronage.[10]

The convention (or curriculum fair, as it is sometimes called) has long been the main access point to homeschool customers, and successful businesses have courted these customers by logging hard hours working booths at conventions around the country and, if possible, winning attendees' trust and affection by lecturing or providing seminars. Evangelical publisher Zondervan's market research found that 49 percent of homeschoolers in 2003 bought their materials at conventions. But in the last few years attendance has begun to decline at many conventions around the country. John Holzmann, whose Sonlight curriculum has long been a convention mainstay, noted that by 2006 many of the venues he and others on his staff frequented were declining. Nevertheless, their business has continued to grow. He hypothesized that the Internet may be "meeting the needs of new homeschoolers in a way that, in years past, only homeschool conventions were able to." Other movement veterans have noticed a similar shift. Susan Richman, editor of the statewide newsletter *PA Homeschoolers*, attributes a marked drop in subscriptions in recent years to the Internet: "As homeschoolers have become more computer savvy, it became easier and easier for local groups ... to create websites and e-lists and discussion groups." As in so many other domains, the Internet has radically altered the way many homeschoolers disseminate and obtain information, making it far more difficult for a few individuals or organizations to define the movement, corral dissent, or control what novices learn, as they could do at conventions.[11]

Closed Communion Ambiguity in a Wired World

The proliferation and democratization of information about homeschooling has left many movement veterans with a sense of misgiving about where their movement is headed. Increasingly conservative Christian leaders worry that the movement is losing its moorings and look back with nostalgia on the days when, in Lisa Guidry's words, "we didn't have email, or yahoogroups.... We had 'phone trees' and tough knees." HSLDA's Chris Klicka spoke for many when he warned in 2006 that "God will continue to bless the homeschool movement—*but only as long as we keep Christ first and foremost.*" Yet at the same time Christian homeschoolers are thrilled at the public acceptance of homeschooling and celebrate the high test scores, college admissions rates, and other markers of worldly success achieved by so many homeschooled kids. Sociologist Christian Smith has noted that evangelical Protestantism tends to be strongest when it looks and acts like mainstream America even as it thinks of itself as battling against it. This is perhaps what is happening with closed communion homeschooling, locked as it is in a "symbiotic relationship" with the educational system against which it has preached for so long but which homeschooling resembles more and more as its institutions mature.[12]

The Home School Legal Defense Association's recent history acutely embodies these ambiguities. By the mid 1990s there was little left to do for an organization devoted to protecting its members' legal rights to homeschool. Flush with cash and with little litigation to spend it on, HSLDA set about expanding its identity. In 1994 it began the "Madison Project," an under-the-radar effort to raise campaign funds for Christian candidates. Members were mailed a card through *The Teaching Home* magazine asking for money for candidates, all Republican, who promised to "abolish the department of education." Nowhere on the card was HSLDA's name mentioned, leading some members to question both the strategy and the very idea of HSLDA moving beyond homeschooling to partisan politics. HSLDA learned from this experience and in subsequent years created separate organizations through which it channeled its explicit political advocacy. The National Center for Home Education and its Congressional Action Program became the base for HSLDA's lobbying efforts on all sorts of issues. The Political

Action Committee was formed to strategize about which issues to engage and how to mobilize resources. Generation Joshua was formed in 2002 to recruit homeschooled teens to "take back the land" by engaging in campaign grunt work like stuffing envelopes, making calls, and anything else needed to get out the Republican vote in tight races. All of these initiatives were financed by HSLDA member contributions. Despite the dwindling legal threats to homeschooling, HSLDA's membership rolls have continued to expand along with the movement as a whole. Though HSLDA loses 18 to 20 percent of its members every year, to date there have always been enough new recruits to more than make up for the losses: around 25 percent of members in a given year are first-timers. Christians new to homeschooling are often understandably worried about doing something so scary, and when they hear over and over from curriculum providers, conference speakers, and support group leaders (many of whom get kickbacks from HSLDA for members they bring in) that they need to join HSLDA, they do. Enough remain to keep the organization solvent, but many new recruits quickly get their bearings and decide to pocket the membership fee after a year or two. So long as the Christian wing of the movement continues to grow at a steady clip, HSLDA will likely do so as well, but should it taper off, HSLDA's coffers will too.[13]

HSLDA's most ambitious and controversial political initiative has been its founding and shepherding of Patrick Henry College. Patrick Henry College (PHC) opened its doors in the fall of 2000 with a class of ninety students, all of them homeschooled, and with Michael Farris as the institution's president. The HSLD Foundation, another HSLDA offshoot, had purchased twenty-nine acres of land in Loudon County, VA, for $400,000 and quickly raised $9 million from parents and big donors, the largest gift coming from Tim LaHaye. A stately building was constructed and the school began with only one major: government. From its inception Patrick Henry has garnered a lot of media attention because of its success in getting its students placed in internships and other openings with key Republican politicians. Of the sixty-four students who graduated in its first four years, twenty secured jobs in Washington. The school made a huge splash in 2004 when its students landed seven of the 100 coveted White House intern positions, an astonishing number given its size. But PHC has also been bedeviled by internal tensions. Farris' confrontational leadership style has alienated many staff, especially faculty. By the end of 2006 seventeen of the twenty full-time faculty members the school had employed in its short history had either been fired or resigned. Most left over issues of academic freedom and theological belief. The college that began by promising to be "Harvard for homeschoolers" has thus far failed to gain accreditation by reputable bodies, two of which have turned it down. In April of 2007 it finally received accreditation from the Transnational Association of Christian Colleges and Schools, an organization that certifies mostly little-known Bible Colleges and missionary training schools. All of this and much else besides have garnered PHC quite a bit of bad publicity, so much so that in 2006 it announced a major restructuring. Farris stepped down from the Presidency to become Chancellor and the school hired a new President and Provost who promised more irenic leadership and greater academic freedom for faculty. Though future years may reverse its fortunes, current alumni of the school are typically reticent about their *alma mater*. Almost all of those who work in Washington have refused to grant permission to PHC's development office to use their pictures in promotional materials, and few will speak on the record about their experiences at the school.[14]

Despite such systemic problems, PHC continues to attract some of the most talented children to emerge out of closed communion homeschooling. Their families are lured to PHC partly because the school is one of the few places in the country that embraces the countercultural natalist vision so many of these families practice. For example, dating is verboten at PHC. "Biblical Courtship" is the norm. At Patrick Henry a young man must secure permission from a young woman's father before pursuing her, and the pursuit must be sexually pure and marriage-driven. Families are also lured to PHC by the promise of political influence and worldly success. Farris has frequently shared a dream he has that one day "an Academy Award winner will walk down the aisle to accept his trophy. On his way, he'll get a cellphone call; it will be the President, who happens to be his old Patrick Henry roommate, calling to congratulate him." Many of Patrick Henry's students come from the American heartland, from families of modest means who are faithful political conservatives. Many of them see PHC as the vehicle by which their children will be transported to careers of influence and success, and they have been primed by PHC to think this way. Robert Stacy, chair of the department of government and one of the school's most compelling personalities until his firing in March of 2006, put the school's pitch this way, "the transition from homeschool to changing the world, leading the nation, shaping the culture, that's us. We are the transition." Parents choose PHC because it promises to protect their kids from the world even as it prepares them to govern it, because it promises to strengthen their relationships with both God and mammon.[15]

Hybridizing the Movement

Ambiguity is to be found not only in the hearts and souls of closed communion homeschoolers. Recent political and legal developments have made it much more difficult to draw sharp distinctions between homeschools and plain old schools. Homeschoolers are increasingly creating hybrids that blend elements of formal schooling into the usual pattern of a mother teaching her own biological children at home. One of the simplest hybrids is the "Mom School." Pioneer Utah homeschooler Joyce Kinmont explains: "a Mom School happens when a mother is homeschooling a child who wants to do something that can be done best in a group, so she invites other homeschooling families to join her. The mom is the teacher." Related but slightly different is the homeschool cooperative, a very popular form of education wherein a group of mothers (and sometimes fathers) pool their expertise, each teaching a subject he or she knows well to all the children in the group. Sometimes such co-ops are held in the homes of respective group members, but often they meet in area churches or other buildings. The most successful and developed of these begin to look quite a bit like schools, sometimes even hiring experts to teach advanced subjects like calculus, foreign language, or physics. Some of these co-ops even *look* like schools to the outside observer, with an adult teacher in the front lecturing to rows of students sitting quietly at desks. Others, however, carry the more free-flowing pedagogy of many homeschoolers into the new setting. North Star, a Massachusetts cooperative that bills itself as "self directed learning for teens," was formed in 1996 by two disgruntled public school teachers and has been strongly influenced by individuals affiliated with John Holt's *Growing Without Schooling*. At North Star no attendance is taken, no grades or evaluations offered. Students learn about whatever they want. In 2006 students asked for and got tutoring in Greek mythology, historical interpretation, Shakespeare, prime numbers, martial arts, culture and belief, electronic music, dance, historical fiction, gaming, ninja science, drawing, makeup technique, Star Trek, theater, writing, and more. Most students engage in apprenticeships and internships in the local community. Though they receive no transcript or degree, to date 49 percent of the alumni have been accepted to college and 49 percent have gotten full- or part-time jobs.[16]

Though disparate in tone and content, such co-ops and other hybrids have emerged at roughly the same time because by the late 1990s and 2000s the homeschooled population was aging. While large numbers of homeschooled kids transition to traditional schools in their teen years, homeschooling through high school has emerged as one of the most popular topics of discussion at conferences and on the Internet in recent years. The number of homeschooled young children is beginning to plateau, but homeschooling for older children is still a high-growth market. Most of the momentum has come from conservative evangelicals, many of whose leaders have lately been expressing great fear that they are losing their teens to secularism. For decades the typical evangelical strategy to combat mass youth culture has been to provide a clean, Christian equivalent. Christian teens have rocked, rapped, moshed, and punked for Jesus for quite some time now. But some Evangelical leaders believe this strategy has proven a failure, and they have been calling instead for a more radical break with the culture. In 1999, after the Monica Lewinsky scandal and the failed effort to remove President Clinton from the White House, veteran political strategist Paul Weyrich published a now famous letter urging Christians "to drop out of this culture, and find places, even if it is where we physically are right now, where we can lead godly, righteous, and sober lives." Pastor E. Ray Moore created "Exodus 2000" (later Exodus Mandate) as a retort to President Clinton's Goals 2000 educational agenda. Moore proclaimed that *"ALL Christians should immediately remove their children from the government schools."* James Dobson picked up on the call and began advocating for the removal of all Christian children from public schools on his radio show in 2002. Other leaders and organizations have added their voices, from D. James Kennedy to Dr. Laura Schlesinger, from the Alliance for the Separation of School and State to a vocal faction within the Southern Baptist Convention who see public schools as the main cause of apostasy among their young. Such calls have brought a fresh batch of formerly public educated teens into to the homeschooling movement. Even as this was happening, thousands of families who had begun homeschooling in the 1980s and early 1990s were trying to figure out what to do with their older kids as well.[17]

The result has been an explosion in innovative programs for older children. "The home-school movement," writes Peter Beinart, "has literally outgrown the home." Beinart found that by 1998 Wichita's 1,500 homeschooling families had created "3 bands, a choir, a bowling group, a math club, a 4-H club, boy-and-girl-scout troops, a debate team, a yearly musical, two libraries and a cap-and-gown graduation." Technically "homeschooled" children were meeting in warehouses or business centers for classes "in algebra, English, science, swimming, accounting, sewing, public speaking, and Tae Kwan Do." The College Board has seen a dramatic rise in homeschoolers who take Advanced

Placement tests. A total of 410 homeschooled students took them in 2000 and 1,282 did so in 2005. Pennsylvania homeschooling veteran Susan Richman reports, "our Advanced Placement online test prep classes have grown tremendously, and are now a major part of our offering." Homeschooling diploma services have multiplied across the country, as have honor societies like the Houston-based Eta Sigma Alpha. Homeschoolers have in recent years challenged, and are increasingly overturning, laws barring them from participation in high school sports and other extracurricular activities. Many school districts, having lost the fight to criminalize homeschooling, now openly court homeschoolers. School districts with high rates of homeschooling have seen significant drops in funding, tied as it is to per-pupil enrollment. The Maricopa County school district in Mesa, AZ, for example, had lost $34 million due to the exodus of 7,526 homeschoolers by the year 2000. In an effort to win some of them back, the district began offering a-la-carte services through satellite campuses at strip malls and other locations. Homeschoolers there have attended weekly enrichment classes in such subjects as sign language, art, karate, and modern dance. The district receives one-fourth of each pupil's government allocation for every student it enrolls in one of the classes.[18]

Not surprisingly, all of this innovation and experimentation at the secondary level, having increased the number of homeschooled high schoolers, has led to a dramatic rise in application for admission to institutions of higher education by students without a traditional high school background. In 1986 ninety percent of the nation's colleges and universities had no explicit homeschooling admissions policy. A recent study found that by 2004 over 75 percent did and that the vast majority of admissions officers surveyed had very positive feelings about homeschooled applicants. In 1998 Congress amended the Higher Education Act to allow homeschooled children to qualify for federal financial aid at colleges and universities by simply self-certifying that they had completed homeschooling in accordance with their state's law (the amendment was drafted by HSLDA's Chris Klicka), making the transition easier for families. Another recent study of the homeschool admissions policies of seventy-two colleges and universities and the subsequent performance by homeschooled students who were enrolled found that homeschoolers were generally happy with the way they were evaluated and universities were happy with the performance and graduation rates of the homeschoolers they admitted. It was all very normal.[19]

New programs blurring the boundaries between home and school have created tensions among some homeschoolers, however. Many with roots in separatist closed communion groups have had trouble adjusting to the more worldly culture that often accompanies cooperatives, especially ambitious endeavors like sporting leagues and orchestras. Disagreements over "what kinds of uniforms are appropriate for home-school cheerleaders and whether rock music may be played at home-school events" are not uncommon as lifelong homeschoolers rub shoulders with each other and with new recruits fresh from the public schools. Not a few Christian homeschoolers have pulled their children out of co-ops, clubs, and sports teams over such concerns. None of these small-scale culture skirmishes begin to match, however, the controversy generated by what is fast becoming the most successful hybrid of all, the cybercharter movement.[20]

Cybercharter schools are only one of many forms of online home-based education that have emerged in recent years. Many public school districts and state educational agencies have been offering online education as a form of distance learning for years. The most innovative and successful of these programs is the Florida Virtual School (FVS), founded in 1997 and operated by the Florida Department of Education. It partners with all sixty-seven Florida school districts to bring a complete high school curriculum moderated by certified teachers to the homes of residents across the state, many of whom live on isolated produce farms or ranches. In the 2005–2006 school year, over 31,000 students were enrolled in FVS. By 2006 twenty-one other states and several local districts had begun similar programs both to service homebound or other special needs students and as an effort to lure homeschoolers (and the tax dollars they represent) back into the public education system.[21]

More controversial have been online schools founded by private companies that have taken advantage of charter school laws in various states to make their services available for free to homeschoolers. The official title for such schools is "nonclassroom-based charters," though they are more often referred to as "cybercharters" or "virtual charters." They are of course only available in states that have passed charter school legislation, but where they are legal they have grown enormously. California was an early innovator in this regard, with virtual charter schools opening shortly after the Charter Schools Act was passed in 1992. By 2001 the state had ninety-three cybercharters serving over 30,000 students, which meant that over 200 million dollars of California's public school budget was being paid to private firms offering homeschool curricula and technology. After it became clear that some of these outfits were making scandalous profits by offering very minimal services, California legislators passed SB 740, which imposed strict financial guidelines on cybercharters, including a requirement that they spend at least 50 percent of public revenues on salaries and benefits to state-certified teachers. The law also set limits on pupil-teacher ratios, required more expansive record-keeping, and imposed strict penalties for failing to meet these and other standards. Over the next few years several California charters failed to meet such requirements and saw their funding cut by 5 to 40 percent. Many did not survive.[22]

By 2006 eighteen states had a combined total of 147 virtual charter schools educating over 65,000 students. Cybercharters in many of these states have faced growing pains similar to those in California. Initial charter school legislation had usually not anticipated the trend toward virtual charters and thus had provided no statutory language to regulate it. After a few years of unbridled innovation and not a little lawless profiteering, most states tightened regulations and increased scrutiny of these programs, causing some of the earliest cybercharters to go out of business. More reputable organizations have prospered, however, leading to conflicts with other public schools. The Western Pennsylvania Cyber Charter School (WPCCS), for example, opened its virtual doors in the fall of 2000 as Pennsylvania's second cybercharter and the first to offer its services across district lines. In its first two months enrollment surged from 250 to over 500. Over half of those enrolled had previously been either homeschooled or had attended private schools. After nine months, enrollment topped 1,100. By 2006 the school had dropped the word "Western" from its name and was employing 400 people to educate 4,400 students on a $30 million budget. Growth of such magnitude, not surprisingly, led to conflict. Many school districts, frustrated that they now had to pay an outside organization to educate students in their own districts, many of whom had not even attended public schools before, simply stopped making payments, causing WPCCS to lose nearly $1 million in 2001. The PA Department of Education responded by withholding $850,000 in state aid from sixty districts, sending the money directly to WPCCS. Lawsuits were filed by twenty-three districts across the state, and in May of 2001 Commonwealth Court senior Judge Warren Morgan ruled against the school districts. That did not stop complaints about WPCCS and several other cybercharters operating in Pennsylvania, however, so in 2002 the state legislature passed Act 88, shifting authorization of cybercharters from local districts to the Department of Education and setting more rigorous requirements and accountability measures. Some of PA's cyberschools, most notably TEACH-Einstein, had their charters revoked, and those that survived tightened their lines considerably, to the great frustration of the many formerly independent homeschoolers lured by free computers and textbooks into these schools but resentful of the regulations and regimentation that seemed to increase every year. Nevertheless, the movement continues to grow. In the 2006–2007 school year the state's eleven cybercharters enrolled about 17,000 students.[23]

Cybercharters have many enemies. Many politicians and public school people agree with Texas Democratic Representative Patrick Rose that "we ought not be in the business of supporting for-profit education. Any program that takes money out of our public schools would be against our better judgment." Profits have indeed been sweet for many cybercharters, operating as they do without extensive facilities and support staff, though advocates always stress the cost of developing curriculum, advertising, and computer networks, as well as the high turnover rate such schools typically endure. They also note that textbook suppliers have been making tremendous profits from public schools for decades. Some cybercharters have been criticized by public school advocates for supporting religious instruction with public monies. In the early years of the movement cybercharters in some states allowed parents to purchase their own curricula, and many chose religious homeschooling materials. One Kansas public school district used the Bob Jones complete curriculum for a short time in the statewide cybercharter it sponsored. Scott Somerville, at the time working for HSLDA, was called by Bob Jones about the propriety of this arrangement. He later reported telling them that since Kansas was one of twelve states with no state constitutional prohibition against spending taxpayer dollars on sectarian education, "who was I to tell BJU to turn down the largest order they'd ever received?" With time, however, state governments have clamped down on such activities. Carol Simpson, Alaska Department of Education program coordinator, noted that when the Interior Distance Education of Alaska program was started in 1997, "we bought nearly anything anyone wanted, including Bob Jones, Alpha Omega, A Beka, etc." A few months later, however, "the Department of Education...made a new regulation prohibiting school districts from purchasing religious curriculum materials." Charter schools are now, with rare exception, wholly secular.[24]

Charterschool advocates and entrepreneurs were not surprised at the criticism (and lawsuits, nearly all of which have been unsuccessful) they have been handed from public school districts, the National Education Association, Democratic legislators resistant to educational choice initiatives, and teachers' unions. What has taken them off guard, however, is the vocal and bitter opposition to the trend among many leaders in the homeschooling community. Closed communion and open communion leaders who had not spoken to one another since the demise of their coalitions in the late 1980s have set aside longstanding feuds and grudges to forge a united protest against virtual schools. In 2003 dozens of homeschool leaders signed a resolution condemning virtual charter schools called "We Stand for Homeschooling." Most of the original signers were open communion veterans (including Linda Dobson, Patrick Farenga, Mark and Helen Hegener, Larry and Susan Kaseman, Mary McCarthy, and Pat Montgomery), but the statement has also been signed by many in the closed communion orbit, including Samuel Blumenfeld, Scott Somerville, Mary Leggewie, and representatives of several Christian support groups and statewide organizations. HSLDA, while not signing the statement, has heaped condemnation on cybercharters. Chris Klicka has called the virtual charter movement a "Trojan horse," warning that "if anything can destroy the Christian homeschool movement, this will." For years now HSLDA has vociferously reiterated that it will terminate the membership of anyone enrolling in a cybercharter. Elizabeth Smith, wife of HSLDA president Michael Smith, urged her audience of closed communion leaders in a 2007 address to pray against the "assault of the Enemy" that cybercharters represented: "if we will band together...through prayer...our groups will win."[25]

Cybercharter backers have had difficulty understanding and responding to such criticism. Ron Packard, founder and CEO of K^{12}, a curriculum provider whose services reached over 25,000 students through virtual schools in sixteen states and the District of Columbia in 2007, has been "shocked" by opposition coming from HSLDA. "It's really amazing to me that a group that has fought so hard for the right to home school would oppose someone else's parents who are fighting for their right to be doing at home a great public school education." Packard noted with some frustration that "the same level of intolerance that you saw in the education establishment toward homeschooling, I think home schooling [groups] are showing toward us." Homeschool movement leaders' reactions do make sense, however. Though the political climate today is quite favorable toward homeschooling, veteran leaders have vivid memories of earlier days when homeschooling was not so easy. Their libertarian rhetoric has demonized government-run schooling for so long that it is very difficult for many of them to think in terms of new paradigms of cooperation and hybridization. Any rapprochement with government is by definition capitulation to the enemy. Animus against government was what bound leftist and conservative Christian homeschoolers together in the late 1970s and early 1980s, and it is what has brought them back together to oppose virtual charters.[26]

Moreover, for veteran Christian leaders especially, the cybercharter movement has cut in on their business. Chris Klicka explains how "even in independent-minded states like Idaho and Alaska" Christian homeschoolers are enrolling in cybercharters "by the thousands. They are attending government homeschool conferences (where Christ or God cannot be mentioned) and receiving the secular, government homeschool newsletters. They no longer go to the Christian homeschool conventions." Christian curriculum providers have lost market share as well. John Holzmann, co-owner of Sonlight Curriculum, notes that "Sonlight, which carries about 95% 'secular' books, has been specifically EXCLUDED as an approved vendor by at least one California Charter School...because of the 'o' in our name: 'They are a Christian company. We cannot support sectarian companies.'" Tens of thousands of Christian homeschoolers who would otherwise be purchasing Christian curricula as independent homeschoolers now receive nonreligious, government sanctioned curriculum for free through cybercharters. It is only natural that Christian companies would resent such a move and interpret it as a clandestine effort by the secularist state to destroy them. Despite the united front of opposition, however, with studies of established virtual charters finding high levels of parent satisfaction and student achievement, it is highly unlikely that independent homeschoolers and advocates for traditional public schools will be able to stop them.[27]

Diversifying the Movement

When virtual charter schools were inaugurated, the great majority of their first clients had been previously homeschooled. By 2006, however, in at least one major California virtual charter, "the majority of families (70%)" were coming from "public and private schools." This new wave of recruits to home-based education has only added to the increasing diversity of people educating their children at home. Though by no means a representative or comprehensive source, it is at least suggestive that by August of 2007 Yahoo!Groups listed 3,930 homeschool discussion groups in the United States alone. There were groups for Catholics (177 of them), Lutherans (6), Muslims (210), Jews (34),

Mormons (83), Jehovah's Witnesses (5), Atheists (35), Pagans (87), gays and lesbians (12), single moms (17), African Americans (26), Native Americans (10), Hispanics (8), and of course hundreds of groups for every conceivable flavor of evangelical or fundamentalist Protestant. With the exception of North Dakota, every state in the union had at least three discussion groups devoted to regional matters, many had dozens, and three (Texas, California, and Florida), had over 100. While terms like "inclusive" and "Christian" still appear with some regularity in the self-description of many groups, suggesting that the old dichotomy between closed and open communion has not gone away, it is clear that a much wider slice of the population teach their children at home now than did in years past.[28]

The increasing participation of African Americans in homeschooling has drawn quite a bit of media attention in recent years, partly because the movement was associated so strongly with conservative WASPs in the 1980s and 1990s, and partly because the vexing achievement gap between black and white students that has inspired so many educational reform initiatives continues unabated. If we think of homeschooling as a deliberate rejection of and alternative to government schooling, then it clearly has a long history among African Americans, going back to the days of slavery when literacy could often be imparted in no other way. Even so, the segregation and lack of school access blacks have endured led many post-emancipation African Americans to look to the school as a "pillar of fire by night after a clouded day," in the words of W.E.B. DuBois. But even as many blacks looked to the schools to bring them freedom and prosperity, others were concluding that public education was part of what was holding them down. Throughout the twentieth century isolated black families kept their children out of public schools and taught them at home. By the early 1990s a few of them began to get organized. One of the first to do so was Donna Nichols-White, who never sent her three children, born in 1986, 1988, and 1993, to school. When Khahil, her oldest, turned five, "it was time to break the news to family, friends, and neighbors that I was going to teach him at home." She and her husband Clifford explained to skeptics their reasons, which included their "happiness with our life as a family, the low academic expectations of the public schools, my distaste for the 'group think' encouraged by the schools, the appalling statistics reported weekly on the failure of Black children in school—regardless of their varying economic circumstances." Always the only African American in support groups or at conventions, Nichols-White "needed to know if there were other families like mine who did what we did, so I started a magazine." *The Drinking Gourd*, named after the famous underground railroad song, was the first homeschooling magazine published by and for minority homeschoolers. It repeatedly emphasized how schools have not helped African Americans and other minorities succeed in America, and how homeschooling offers an escape from this cycle of failure. The magazine was ahead of its time and was only published for a few years in the 1990s, but it did bring small groups of minority homeschoolers together for the first time, providing a forum for them to share their experiences. Moreover, it put the issue of minority homeschooling on the radar screen for the larger movement.[29]

Some statewide organizations began holding a session or two on minority homeschooling at their annual conventions in the 1990s. Year by year more African Americans attended. Longtime homeschooling observer Brian Ray noted in 2006, "in the 1990s, you saw a little more color, and by 2000, a substantial number of black families started showing up. In some cities, the majority of those attending conferences are African American." Even so, attending homeschooling conventions and support groups could still be a harrowing experience for an African American family in 2000. Christian history curricula very often celebrated the "Christian" America of the colonial, early national, and antebellum periods, and not infrequently described the confederate South in glowing terms. European, or "Western," civilization was stressed in many popular curricula, and illustrations in many of the most popular homeschooling textbooks rarely depicted minorities. Historian George Marsden has noted that the emphasis the religious right places on the United States having historically been a Christian nation is the main reason African Americans, many of whom share the religious and cultural attitudes of white conservatives, are distrustful of the movement. They remember the slavery and oppression antebellum Christians sanctioned and the racism that white Christians possessed in spades after that.[30]

Nevertheless, homeschooling among African Americans has grown rapidly since the late 1990s. The U.S. Department of Education estimated that by 2003 there were 103,000 Black homeschoolers. Nonprofits like the Children's Scholarship Fund, founded in 1998, have provided vouchers to help low-income families attend private schools, and some are using the money to homeschool. Several support groups have formed to build momentum: In 1996 Gilbert and Gloria Wilkerson created the Network of Black Homeschoolers to provide support and networking among its clientele. By 2002 the organization had 300 members. In 2000 Joyce and Eric Burges created the National

Black Home Educators Resource Association to offer curriculum advice, pair newbies with veterans, and organize an annual symposium. By 2005 the organization had a mailing list of 2000 families. The Burgeses and the Wilkersons were featured in HSLDA's *Home School Court Report* in July of 2001, a landmark event in HSLDA's self-image. Prior to this issue there had never been a minority face on a cover of the *Court Report*. Thereafter, however, minorities were consistently represented. HSLDA has committed itself to encouraging the growth and visibility of minority homeschooling on many fronts. It has financed conferences for the Burges' organization, sponsored research on black homeschooler achievement, and recruited heavily (with very little success) from the African American homeschool community to diversify the student body at Patrick Henry. Sara Diamond, a careful student and sharp critic of the religious right, has noted that despite her many reservations, "the most active efforts at racial integration are underway within conservative denominations, not within the liberal churches." HSLDA has taken a strong stand here even though it notes that some of its own members "still seem to be prejudiced."[31]

The newest and largest of the national homeschool organizations for blacks is the National African-American Homeschoolers Alliance, founded in 2003 by Jennifer James. By 2006 the organization had 3,000 members. James learned of homeschooling by watching the success of homeschoolers at the Scripps National Spelling Bee and embraced it for her own family. She quickly felt isolated, however, "We hadn't heard or read about, or run into, or known any black homeschoolers other than ourselves." Like the Burgeses and many other black families, she encountered strong opposition from many in her community who saw it "as a betrayal of the Civil Rights movement." But she has been winning these people over, frequently citing a 1997 study by Brian Ray's National Home Education Research Institute (NHERI) that found no achievement gap between black and white homeschoolers based on results from the Iowa Assessment Test. Ray found that black homeschoolers scored in the 87th percentile in reading and 77th percentile in math, far above the level of schooled blacks and on par with whites. Though the study, as with many of NHERI's efforts over the years, has serious limitations from a social-scientific standpoint, it was greeted with aplomb by James and others in the black press for its message that African Americans can do just as well as whites. "Families are running out of options," James notes, "There's this persistent achievement gap, and a lot of black children are doing so poorly in traditional schools that parents are looking for alternatives." Homeschooling is becoming the option of choice for many, and as such "the Black homeschool movement is growing at a faster rate than the general homeschool population," in HSLDA's Michael Smith's words. Many locales have experienced enough growth in recent years to sustain "African-American specific" support groups. In many cases these groups are every bit as religious as closed communion groups in the white community, and most black homeschoolers are pedagogical traditionalists as well. Diversity is increasing here too, though, as illustrated by *FUNgasa*, a glossy online magazine debuting in 2004 by the group African-American Unschooling, founded by S. Courtney Walton.[32]

While the recent growth of homeschooling among African Americans has attracted the most press, the trend can be spotted among many other groups as well. Native American homeschool organizations have been founded to escape assimilationist public schools and preserve Native values in Virginia and North Carolina. Similarly, many Hawaiian Natives have found homeschooling to be the solution to the gulf between tribal ways and public education. Jews, especially the Orthodox, have been homeschooling in much greater numbers in recent years. While Roman Catholic families have long had a presence in the homeschooling world with such institutions as the Virginia-based Seton Home Study School (founded in 1980), recent years have seen an explosion in Catholic homeschooling and resources. Islamic homeschooling has also grown rapidly, especially since 9/11, largely because "the public school system is not accommodating to Muslims," in the words of Fatima Saleem, founder of the Palmetto Muslim Homeschool Resource Network. Homeschooling has become quite popular among neo-Pagans, Wiccans, and other adherents of alternative religions because it allows them to escape from schools they see as "embodying secular science's rationalized world view" and to impart concepts of "individual potential, spirituality, and holism" to their children. Many Aryan Nations' members and other white nationalist "Folk" are strong advocates of homeschooling and have a robust presence on the Internet. Large numbers of parents whose children have been diagnosed with learning disabilities have pulled them from schools, believing they can do a better job teaching them at home.[33]

Increasing numbers of wealthy Americans are hiring private tutors. The U.S. Department of Education estimated that 21 percent of homeschoolers were being taught this way in 2003. The largest tutor-supply company, Professional Tutors of America, has 6,000 tutors on its payroll but still cannot meet even a third of the in-home requests it receives. Michelle Conlin has explained the appeal of home education to "creative-class parents" as an outgrowth of the "spread of the postgeographic workstyle" and "flextime economy." Home schooling "can untether families from zip codes and school districts" even as it prepares children for "the global knowledge economy."[34]

A final group of homeschoolers that should be mentioned in this brief catalogue of diversity is the conglomerate of children who engage in one form or another of intensive extracurricular activity. Children involved in sports requiring rigorous training, acting and modeling, demanding arts or music programs, and so forth are often homeschooled. In motocross, where an elite-level thirteen-year-old can earn over $100,000 a year, 90 percent of minors are either homeschoolers or dropouts. Circe Wallace, a retired snowboarder turned action-sport agent, remarked in 2006, "I've been in this business 15 years, and it's always been those with parents that understand the freedom and flexibility of home-schooling that go the furthest." Orange County gymnast Katy Nogaki was eleven years old when she told a reporter, "my coaches... said if I home-schooled, I could come to the gym early and I could get really far in gymnastics.... When I was in regular school, I wasn't as good, but when I was home-schooled, I got state champion." Actress and singer Vanessa Hudgens, like many other kids trying to make it in Hollywood, "didn't really have any time to go to a real school." Her mother homeschooled her from the 8th grade on between acting classes and auditions. The family's sacrifices were abundantly rewarded when *High School Musical* and its sequel became worldwide sensations, making Hudgens a household name.[35]

In short, home education is now being done by so many different kinds of people for so many different reasons that it no longer makes much sense to speak of it as a movement or even a set of movements. For an increasing number of Americans, it's just one option among many to consider, for a few months or for a lifetime. Yet many of the new breed of home schooling parents still need help with difficult pedagogical or curricular decisions, playmates for their children and companionship for themselves, and opportunities just to get out of the house for a while. Home school support groups today can be some of the most diverse conglomerates in the country, as Pam Sorooshian's description of her Southern California group attests:

> My homeschooling group includes Moslem, Jewish, Quaker, Baptist, Messianic Jews, Pagan, Baha'i, atheist, agnostic, Catholic, unity, evangelicals, other protestant denominations, and probably more. We have

African Americans, Latinos, Asians, Middle Easterners, and other minorities. We have stay-at-home dads and single mothers. We are FAR more diverse than the neighborhood school I pulled my oldest child out of 10 years ago.[36]

The Meaning of the Movement

The growing appeal of homeschooling to all sorts of people has led to many, often contradictory, claims about its broader significance. Colleen McDannell, for example, agrees with several in the Christian wing of the movement who interpret homeschooling as a revival on par with some of the greatest awakenings of American history, but she notes that this revival is not so much transforming American society as it is transforming conservative Protestantism itself, especially its understanding of gender. Homeschooling mothers "no longer see themselves as simply housewives or mothers." Home becomes workplace; the mother an educational professional. Fathers are urged to become more domesticated. Boys learn to cook, clean, and take care of younger siblings. Children in general are raised with less gender specificity. Other scholars studying homeschooling have noted how a movement "generated partly in reaction to feminism" has nevertheless selectively incorporated "many feminist family forms," including the softening and domestication of the male, the therapeutic orientation to marriage and child-rearing, and of course the provision of excellent education to girls. Women form "the backbone of the homeschool movement's impressive organizational system," empowered by their belief in a God-given vocation to live a life of powerful dissent from established norms even as they try to convince others that homeschooling is, after all, pretty normal.[37]

What some scholars see as feminism, however, others see as an example of antimodernism. For some commentators, the phenomenon of mothers educating their children at home is no feminist trope but a compelling attempt to bypass "the evasive banality of modern culture" and pass on to children a hopeful vision of transcendence, of faith that there is more to life than the transient thrills of mass market consumerism. Allan Carlson, a thoughtful and prolific family advocate and critic of modernity, understands homeschooling to be part of a larger trend of "deindustrialism," which, along with home-based businesses and small-scale communities, represents a move among some in postindustrial America to rethink patterns of living that have been with us since the Civil War. Taking the long view, feminism was simply a reaction to modernist gender dichotomies whose deepest sources were more economic than cultural: dad goes off to make money and mom stays home to spend it. Homeschooling takes the critique to the root cause: an industrialism that has given us not only gender dichotomies but vastly reduced birth rates and a depopulated countryside. Homeschooling parents, with their high birth rate, penchant for gardens and livestock, and willingness to sacrifice individual self-actualization for intensive child-rearing, represent for Carlson and some other scholars one of the few positive signs of resistance to the secularism and population decline currently destroying wealthy societies from within.[38]

As we have seen, however, most homeschoolers do not live on forty acres, nor do most have nine children (though thy do have larger families on average than do Americans who don't homeschool). Another set of observers finds in homeschooling not an alternative to modernity but a pure, perhaps the most pure, expression of it. Polish born sociologist Zygmunt Bauman has recently written a string of books explicating what he calls "liquid modernity," the cultural condition where each individual is free to do as he or she pleases in an endless and boundless present but absent compelling social norms that would make such choices meaningful. For many commentators, homeschooling is part of this move to liquidity, as the grounded, industrial-era education of the past gives way to a cybertopian form of libertarian education. In this telling, homeschoolers, for all the social capital they amass through myriad support groups and organizations, are at heart deregulators, interested in maintaining their own autonomy and independence from government or any other force that would impose limits on them. Pamela Ann Moss has shown that while homeschoolers think that society as a whole will eventually benefit from the strong families that homeschooling helps create, what they really want is, in Chris Klicka's words, "the right to be left alone." Many of the homeschooling movement's most articulate critics (notably Michael Apple, Rob Reich, and Chris Lubienski) charge that homeschooling is essentially an exercise in evasion of difference, an extreme form of the "secession of the successful" from meaningful engagement with public life. At the same time, a number of futurists celebrate homeschooling for this very thing. As the Internet has expanded the purchasing and entertainment choices people have, replacing the old days of mass consumption with unlimited niche markets, as America's "new independent workers are transforming the way we live" by erasing the boundaries of geography and company loyalty through their free agency, so homeschooling represents the future of education: deregulated, market-driven, privatized, malleable, liquid.[39]

Feminist, antimodernist, libertarian, or, as is likely the case, a little of all of them plus much else besides—in as many combinations as there are homeschoolers—whatever be the meaning of homeschooling, at least two things can be said for it today. First, homeschooling is not going away, as some public school officials believed it would in the early 1980s. Second, "homeschooling" has now become "home schooling," the use of the home to teach children for any number of reasons. Many, perhaps most, who do so still choose this option out of frustration with or protest against formal, institutional schooling. They are still homeschoolers. Others do so as an accessory, hybrid, temporary stopgap, or out of necessity given their circumstances. They have become the new domestic educators. Whether the movement as a whole will ultimately be seen as a countercultural protest, an embodiment of the *zeitgeist*, or perhaps both at the same time, cannot be predicted with certainty. What we can say is that with the legal ground cleared, pedagogical and curricular pathways paved and clearly lit, and settlement now dense and growing "by both conversion and conception," there is certainly a bright future for home-based education in the United States.[40]

Notes

Introduction

1. Theodore Forstmann, "Putting Parents in Charge," *First Things* 115 (August/September 2001): 22.
2. For examples of the ahistorical appropriation of historic notables to the cause, see Mac Plent, *An "A" in Life: Famous Home Schoolers* (Farmingdale: Unschoolers Network, 1999); John Whitehead and Wendell Bird, *Home Education and Constitutional Liberties* (Wheaton, IL: Crossway, 1984), 22–25; Linda Dobson, *Homeschoolers' Success Stories: 15 Adults and 12 Young People Share the Impact that Homeschooling Has Made on their Lives* (Roseville, CA: Prima, 2000), 7–24; and Olivia C. Loria, "Twenty-One Years on the Front Lines of Home Education Nationally and Internationally," *Tamariki School* www.tamariki.school.nz/talks/Olivia%20Loria.doc (28 January 2008). Lawrence Cremin, ed., *The Republic and the School: Horace Mann on the Education of Free Men* (New York: Teachers College, 1957), 80.
3. For a detailed history of several of the great figures, see Cheryl Lindsey Seelhoff, "A Homeschooler's History of Homeschooling, Part I," *Gentle Spirit Magazine* 6, no. 9 (January 2000): 32–44, and "A Homeschooler's History of Homeschooling, Part II," *Gentle Spirit Magazine* 6, no. 10 (April/May 2000): 66–70.
4. On recent discussions of historical synthesis, see Ian Tyrrell, "The Great Historical Jeremiad: The Problem of Specialization in American Historiography," *The History Teacher* 33, no. 3 (May 2000): 371–393 and Allan Megill, "Fragmentation and the Future of Historiography," *American Historical Review* 96, no. 3 (June 1991): 693–698.
5. John Demos, *A Little Commonwealth: Family Life in Plymouth Colony* (New York: Oxford University Press, 1970), 183. On house churches, see David Van Biema and Rita Healy, "There's No Pulpit Like Home," *Time* (March 6, 2006): 46–48. On nannies, see Caitlin Flanagan, "How Serfdom Saved the Women's Movement," *Atlantic* (March 2004): 109–128. On homebirth, see Pamela E. Klassen, *Blessed Events: Religion and Home Birth in America* (Princeton: Princeton University Press, 2001). On home care, see Lisa Belkin, "A Virtual Retirement Home," *New York Times Magazine* (March 5, 2006).
6. On complexity in history, see Lawrence W. Levine, "The Unpredictable Past: Reflections on Recent American Historiography," *American Historical Review* 94, no. 3 (June 1989): 671–679. On the emergence and dominance of microhistory see Georg G. Iggers, *Historiography in the*

Twentieth Century: From Scientific Objectivity to the Postmodern Challenge (Hanover, NH: Wesleyan University Press, 1997).

Chapter One The Family State, 1600–1776

1. My account in this and the following paragraphs draws on Francis Dillon, *The Pilgrims* (New York: Doubleday, 1973) and Dewey D. Wallace, *The Pilgrims* (Wilmington, NC: McGrath, 1977).
2. Dillon, *Pilgrims*, 92.
3. William Bradford, *Of Plymouth Plantation, 1620–1647* (New York: Random House, 1952), 17, 24, 25. I have modernized the spelling of this and other colonial sources cited in this chapter for sake of clarity.
4. Demos, *Little Commonwealth*, 144. Arthur W. Calhoun, *Social History of the American Family: From Colonial Times to the Present*, vol. 1, Colonial Period (New York: Barnes and Noble, 1945), 72.
5. Steven Mintz and Susan Kellog, *Domestic Revolutions: A Social History of American Family Life* (New York: Free Press, 1988), 26–31. Stephanie Coontz, *The Social Origins of Private Life: A History of American Families, 1600–1900* (New York: Verso, 1988), 56, 45.
6. Linda Clemmons, "'We Find it a Difficult Work': Educating Dakota Children in Missionary Homes, 1835–1862," *American Indian Quarterly* 24, no. 4 (Fall 2000): 98. Bernd C. Peyer, *The Tutor'd Mind: Indian Missionary-Writers in Antebellum America* (Amherst: University of Massachusetts Press, 1997), 30.
7. Both Rudy Ray Seward, *The American Family: A Demographic History* (Beverly Hills, CA: Sage Publications, 1978), 21–27, 41–45 and Maris Vinovskis, "Family and Schooling in Colonial and Nineteenth Century America," *Journal of Family History* 12 (1987): 19–37 summarize this scholarly debate.
8. Seward, *American Family*, 45–49; Demos, *Little Commonwealth*, 66–70; Lawrence Cremin, *American Education: The Colonial Experience, 1607–1783* (New York: Harper, 1970), 481–483, 128. Mintz and Kellogg, *Domestic Revolutions*, 16. John Demos, "Notes of Life in Plymouth Colony," *William and Mary Quarterly* 3, no. 22 (April 1965): 285. Benjamin Franklin, *Autobiography of Benjamin Franklin* (New York: Dover, 1996), 5.
9. Demos, *Little Commonwealth*, 49–51. Cremin, *Colonial Experience*, 128.
10. Cremin, *Colonial Experience*, 124–125.
11. Cremin, *Colonial Experience*, 125–126.
12. Demos, *Little Commonwealth*, 183–184, 81. Mimi Abramovitz, *Regulating the Lives of Women: Social Welfare Policy from Colonial Times to the Present* (Boston: South End, 1988), 53–54. Mary Beth Norton, *Founding Mothers and Fathers: Gendered Power and the*

Forming of American Society (New York: Random House, 1996), 40–41.
13. LeRoy Ashby, *Endangered Children: Dependency, Neglect, and Abuse in American History* (New York: Twayne, 1997), 12. Steven Mintz, "Regulating the American Family," *Journal of Family History* 14, no. 4 (1989): 390. Calhoun, *Social History of the American Family*, 73. Cotton, quoted in Thomas Hine, *The Rise and Fall of the American Teenager* (New York: Avon Books, 1999), 65.
14. Norton, *Founding Mothers*, 4–8.
15. Calhoun, *Social History of the American Family*, 74.
16. Mintz and Kellog, *Domestic Revolutions*, 54. N. Ray Hiner, "The Cry of Sodom Enquired Into: Educational Analysis in Seventeenth-Century New England," *History of Education Quarterly* 13, no. 1 (Spring 1973): 13. Maris Vinovskis and Gerald F. Moran, "The Great Care of Godly Parents: Early Childhood in Puritan New England" in John Hagen and Alice Smuts, eds., *History and Research in Child Development: In Celebration of the Fiftieth Anniversary of the Society* (Chicago: University of Chicago Press, 1986), 24–37.
17. Marylynn Salmon, *Women and the Law of Property in Early America* (Chapel Hill: University of North Carolina Press, 1986), 6–9. Norton, *Founding Mothers*, 42–50. Demos, *Little Commonwealth*, 54–56.
18. Norton, *Founding Mothers*, 105.
19. Gerald F. Moran, "'Sisters' in Christ: Women and the Church in Seventeenth-Century New England" in Paul S. Boyer and Janet Wilson James, ed., *Women in American Religion* (Philadelphia: University of Pennsylvania Press, 1980), 47–65. Donald M. Scott and Bernard Wishy, *America's Families: A Documentary History* (New York: Harper and Row, 1982), 8. Robert V. Wells, "Family History and Demographic Transition," *Journal of Social History* 9, no. 1 (Fall 1975): 1–19.
20. Gerald F. Moran and Maris A. Vinovskis, *Religion, Family, and the Life Course: Explorations in the Social History of Early America* (Ann Arbor: University of Michigan Press, 1992).
21. Alice Morse Earle, *Home Life in Colonial Days* (New York: Macmillan, 1898), 266–267. Cremin, *Colonial Experience*, 482.
22. Cremin, *Colonial Experience*, 130.
23. Michael Wigglesworth, *The Day of Doom* (New York: Russell and Russell, 1966), 17. Mel Yazawa, ed., *The Diary and Life of Samuel Sewall* (Boston: Bedford Books, 1998), 143, 146, 156.
24. Michael G. Hall, ed., "The Autobiography of Increase Mather," *American Antiquarian Society Proceedings* 71, no. 2 (1962): 278. Vinovskis, "Family and Schooling in Colonial and Nineteenth Century America," 23.
25. Cremin, *Colonial Experience*, 128–129.
26. Barbara Beatty, *Preschool Education in America: The Culture of Young Children from the Colonial Era to the Present* (New Haven: Yale University Press, 1995), 8–9. Edward E. Gordon and Elaine H. Gordon, *Centuries of Tutoring: A History of Alternative Education in America*

and Western Europe (Lanham, MD: University Press of America, 1990), 246, 249. Huey B. Long, "Adult Basic Education in Colonial America," *Adult Literacy and Basic Education* 7, no. 2 (1983): 66–67.
27. Cremin, *Colonial Experience*, 114–115. Demos, *Little Commonwealth*, 71–75.
28. Coontz, *Social Origins of Private life*, 31, 86. Edmund Morgan, *The Puritan Family: Religion and Domestic Relations in Seventeenth-Century New England* (New York: Harper and Row, 1966), 76–78. Williams, cited in Helena M. Wall, *Fierce Communion: Family and Community in Early America* (Cambridge: Harvard, 1990), 105.
29. Wall, *Fierce Communion*, 118, 114–116, 97–110.
30. Stephen M. Frank, *Life with Father* (Baltimore, MD: Johns Hopkins University Press, 1998), 143–144. Wall, *Fierce Communion*, 116–118, 123–124.
31. Wall, *Fierce Communion*, 118–119, 94–95. Ashby, *Endangered Children*, 11, 14. Elizabeth Pleck, *Domestic Tyranny: The Making of Social Policy against Family Violence from Colonial Times to the Present* (Urbana: University of Illinois Press, 2004), 17–68. Norton, *Founding Mothers*, 118–120.
32. John Robinson, "Of Children and their Education" in *The Works of John Robinson*, vol. 1 (London: John Snow, 1851), 246. Philip Grevin's work especially has both clarified and condemned the will-breaking tradition. See his *The Protestant Temperament: Patterns of Child-Rearing, Religious Experience, and the Self in Early America* (New York: Knopf, 1977), 21–150 and *Spare the Child: The Religious Roots of Punishment and the Psychological Impact of Physical Abuse* (New York: Knopf, 1991). For a less judgmental, more historicist view, see Rodney Hessinger, "Problems and Promises: Colonial American Child Rearing and Modernization Theory," *Journal of Family History* 21 (April 1996): 137 and Daniel Scott Smith, "Child-naming Practices, Kinship Ties, and Change in Family Attitudes in Hingham, Massachusetts, 1641–1880," *Journal of Social History* 18 (1985): 541–566.
33. Sewall, cited in Calhoun, *Social History of the American Family*, 80. Cremin, *Colonial Experience*, 485. For a rich description of the world of several sorts of slave children, see Wilma King, *Stolen Childhoods: Slave Youth in Nineteenth Century America* (Bloomington: Indiana University Press, 1995).
34. Carol Berkin, *First Generations: Women in Colonial America* (New York: Hill and Wang, 1997), 132–133, 146–147.
35. Gordon, *Centuries of Tutoring*, 251–269. Salmon, *Women and the Law of Property*, 9–11. Catherine Clinton, *The Plantation Mistress: Woman's World in the Old South* (New York: Pantheon, 1982), especially chapter 7.
36. Brenda Stevenson, *Life in Black and White: Family and Community in the Slave South* (New York: Oxford, 1996), 277, 113.
37. Jay Fleigelman, *Prodigals & Pilgrims* (New York: Cambridge University Press, 1982), 39–40.

Chapter Two The Family Nation, 1776–1860

1. David McCullough, *John Adams* (New York: Simon and Schuster, 2001), 55–58. Phyllis Lee Levin, *Abigal Adams: A Biography* (New York: St. Martin's Press, 1987), 39–45.
2. Henry Adams, *The Education of Henry Adams* (Boston: Houghton Mifflin, 1918), 13.
3. Bernard Wishy, *The Child and the Republic: The Dawn of Modern American Child Nurture* (Philadelphia: University of Pennsylvania Press, 1967), 11–12.
4. For a good introduction to Calvinism, see John Thomas McNeill, *The History and Character of Calvinism* (New York: Oxford University Press, 1967).
5. Fleigelman, *Prodigals & Pilgrims*, 167–168. Margaret Bendroth, "Children of Adam, Children of God: Christian Nurture in Early Nineteenth-century America," *Theology Today*, 56, no. 4 (January 2000): 495–505.
6. Fleigelman, *Prodigals and Pilgrims*, 160, 170–171.
7. Nathan Hatch, *The Democratization of American Christianity* (New Haven: Yale University Press, 1990). Jon Butler, *Awash in a Sea of Faith: Christianizing the American People* (Cambridge: Harvard University Press, 1990). Mark A. Noll, *A History of Christianity in the United States and Canada* (Grand Rapids: Eerdmans, 1992), 173, 178. Ann Braude, "Women's History *Is* American Religious History" in Thomas A. Tweed, ed., *Retelling U. S. Religious History* (Berkeley: University of California Press, 1997), 87. Trollope, cited in Coontz, *Social Origins of Private Life*, 88.
8. Bendroth, "Children of Adam," 502–503. Fleigelman, *Prodigals and Pilgrims*, 186. A. Anne L. Kuhn, *The Mother's Role in Childhood Education: New England Concepts, 1830–1860* (New Haven: Yale University Press, 1947), 81.
9. Charles Sellers, *The Market Revolution: Jacksonian America, 1815–1846* (New York: Oxford University Press, 1991). Coontz, *Social Origins of Private Life*, 167, 29. Mintz and Kellogg, *Domestic Revolutions*, 22, 49–52.
10. Hal S. Barron, *Those Who Stayed Behind: Rural Society in Nineteenth Century New England* (Cambridge: Cambridge University Press, 1984). Alan Kulikoff, *The Agrarian Origins of American Capitalism* (Charlottesville: University of Virginia Press, 1992).
11. Coontz, *Social Origins of Private Life*, 141.
12. Webster and Beecher, cited in Kuhn, *Mother's Role*, 34, 181–182. Alexis de Tocqueville, *Democracy in America*, vol. 2 (New York: Vintage, 1990), 201, 214.
13. Coontz, *Social Origins of Private Life*, 215. Daniel Blake Smith, *Inside the Great House: Planter Family Life in Eighteenth-Century*

Chesapeake Society (Ithaca: Cornell University Press, 1986). Emerson, cited in Richard T. Gill, *Posterity Lost: Progress, Ideology, and the Decline of the American Family* (Lanham: Rowman and Littlefield, 1997), 17–18. Ann Douglas, *The Feminization of American Culture* (New York: Knopf, 1977). Ted Ownby, *Family Men: Middle Class Fatherhood in Early Industrial America* (London: Routledge, 2001). Karen Halttunen, "Humanitarianism and the Pornography of Pain in Anglo-American Culture," *American Historical Review* 100, no. 2 (April 1995): 303–334.

14. Steven Mintz, *A Prison of Expectations: The Family in Victorian Culture* (New York: New York University Press, 1983), 11–39. Elizabeth H. Pleck, *Celebrating the Family: Ethnicity, Consumer Culture, and Family Ritual* (Cambridge: Harvard University Press, 2000). Penne L. Restad, *Christmas in America: A History* (New York: Oxford University Press, 1995).

15. Fleigelman, *Prodigals and Pilgrims*, 28–29, 83, 84. Halttunen, "Humanitarianism," 304. William Charvat, et al., eds., *The Centenary Edition of the Works of Nathaniel Hawthorne*, vol. 17 (Columbus: Ohio State University Press, 1962), 70. Frank, *Life with Father*, 26. Kuhn, *Mother's Role*, 42–46.

16. Solomon Stoddard, "The Way for a People to Live Long in the Land" (Boston: Allen and Green, 1703) Early American Imprints no. 1148. Cotton Mather, *Magnalia Christi Americana*, vol. 1 (Hartford: Silas Andrus, 1820), 59. Kuhn, *Mother's Role*, 72.

17. Beatty, *Preschool Education in America*, 31–37. John Rowe Townsend, *Written for Children: An Outline of English-Language Children's Literature* (New York: Lippincott, 1983), 36–55. Wishy, *Child and the Republic*, 52–54.

18. Kuhn, *Mother's Role*, 181. Kathryn Kish Slkar, *Catharine Beecher: A Study in American Domesticity* (New York: Norton, 1976), 151–167.

19. Carl F. Kaestle, *Pillars of the Republic: Common Schools and American Society, 1780–1860* (New York: Hill and Wang, 1983), 70–72.

20. Kaestle, *Pillars of the Republic*, 75–103. Timothy L. Smith, "Protestant Schooling and American Nationality, 1800–1850," *Journal of American History* 53, no. 4 (March 1967): 679. Charles I. Foster, *An Errand of Mercy: The Evangelical United Front, 1790–1837* (Chapel Hill: University of North Carolina Press, 1960).

21. Frank, *Life with Father*, 150. Geraldine J. Clifford, "Home and School in Nineteenth Century America: Some Personal—History Reports from the United States," *History of Education Quarterly* 18, no. 1 (Spring 1978): 11–12.

22. Barbara Finkelstein, "'In Fear of Childhood': Relationships between Parents and Teachers in Popular Primary Schools in the Nineteenth Century," *History of Education Quarterly* 3 (Winter 1976): 324, 329. Marilyn Schultz Blackwell, "The Politics of Motherhood: Clarina

Howard Nichols and School Suffrage," *New England Quarterly* 78, no. 4. (December 2005): 584.
23. Kaestle, *Pillars of the Republic*, 63. Finkelstein, "In Fear of Childhood," 329, 326, 325.
24. Lawrence Cremin, *American Education: The National Experience, 1783-1876* (New York: Harper and Row, 1980), 177-178. Joel Perlmann and Robert A. Margo, *Women's Work? American Schoolteachers, 1650-1920* (Chicago: University of Chicago Press, 2001), 86-124.
25. Samuel Goodrich, *Fireside Education* (New York: Huntington, 1838), 64. Jacqueline Jones, *Soldiers of Light and Love: Northern Teachers and Georgia Blacks, 1865-1873* (Chapel Hill: University of North Carolina Press, 1980). Polly Welts Kaufman, *Women Teachers on the Frontier* (New Haven: Yale University Press, 1984), xxii.
26. Wishy, *Child and the Republic*, 75. Ruth Miller Elson, *Guardians of Tradition: American Schoolbooks of the Nineteenth Century* (Lincoln: University of Nebraska Press, 1964), 378.
27. Anne M. Boylan, *Sunday School: The Formation of an American Institution, 1790-1880* (New Haven: Yale University Press, 1990), 59. Beatty, *Preschool Education in America*, 38.
28. James W. Fraser, *Between Church and State: Religion and Public Education in a Multicultural Society* (New York: St. Martin's, 1999), 49-66.
29. Geraldine Youcha, *Minding the Children: Child Care in America from Colonial Times to the Present* (New York: Scribner, 1995), 67-114. David Tyack and Elisabeth Hansot, *Learning Together: A History of Coeducation in American Public Schools* (New Haven: Yale University Press, 1990), 46.
30. Margaret Nash, "'A Triumph of Reason': Female Education in Academies in the New Republic" in Nancy Beadie and Kim Tolley, eds., *Chartered Schools: Two Hundred Years of Independent Academies in the United States, 1727-1925* (New York: Routledge Falmer, 2002), 64-66. Gordon, *Centuries of Tutoring*, 281.
31. Jonathan Messerli, *Horace Mann: A Biography* (New York: Knopf, 1972), 429.
32. Selwyn K. Troen, *The Public and the Schools: Shaping the St. Louis System* (Columbia: University of Missouri Press, 1975). Anya Jabour, "Grown Girls, Highly Cultivated: Female Education in an Antebellum Southern Family," *Journal of Southern History* 64, no. 1 (February 1998): 23-64.
33. Jim Powell, "The Education of Thomas Edison," *The Freeman* 45, no. 2 (February 1995), 73-76.
34. Elliott West, *Growing Up with the Country: Childhood on the Far Western Frontier* (Albuquerque: University of New Mexico Press, 1989), xi. Frank, *Life with Father*, 68.

35. Frances E. Willard, *Glimpses of Fifty Years: The Autobiography of an American Woman* (Chicago: Woman's Temperance Publication Association, 1889), 73-97.
36. Gordon, *Centuries of Tutoring*, 289, 292. Julie Roy Jeffrey, *Frontier Women: The Trans-Mississippi West, 1840-1880* (New York: Hill and Wang, 1979), 87-90. Edward E. Gordon, "An Eternity Job: Riding the Literacy Circuit on the Western Frontier" (paper presented at the History of Education Society annual meeting, Evanston, Ill, November, 2003). Terje Ann Hanson, *Home Schooling in Alaska: Extreme Experiments in Home Education* (M.A. diss., University of Alaska Fairbanks, 2000).
37. Ashby, *Endangered Children*, 28.
38. Clemmons, "We Find It a Difficult Work," 95-182.
39. Heather Andrea Williams, *Self-Taught: African American Education in Slavery and Freedom* (Chapel Hill: University of North Carolina Press, 2005), 18-29.

Chapter Three   The Eclipse of the Fireside, 1865-1930

1. Walter Sullivan, *The War the Women Lived: Female Voices from the Confederate South* (Nashville: J. S. Sanders, 1995), 107-111.
2. Sullivan, *War the Women Lived*, 283-291. Clara Junker, "Women at War: The Civil War Diaries of Floride Clemson and Cornelia Peake McDonald," *Southern Quarterly* 42, no. 4 (Summer 2004): 90-106.
3. Laura F. Edwards, *Scarlett Doesn't Live Here Anymore: Southern Women in the Civil War Era* (Urbana: University of Illinois Press, 2000), 91-92, 100-109. James Marten, *The Children's Civil War* (Chapel Hill: University of North Carolina Press, 1998), 203, 231-232. O. H. Oldroyd, *The Good Old Songs We Used to Sing, '61 to '65* (Washington, D.C.: O. H. Oldroyd, 1902), 24.
4. Sullivan, *War the Women Lived*, 289. Mark A. Noll, *The Civil War as a Theological Crisis* (Chapel Hill: University of North Carolina Press, 2006), 31-50, 75-94. Noll, *A History of Christianity*, 329-330.
5. Wishy, *Child and the Republic*, 81-93. Louise Stevenson, *The Victorian Homefront: American Thought and Culture, 1860-1880* (New York: Twayne, 1991), 96-97. George Marsden, *Fundamentalism and American Culture:The Shaping of Twentieth Century Evangelicalism, 1870-1925* (New York: Oxford University Press, 1980).
6. Robert Wiebe, *The Search for Order, 1877-1920* (New York: Hill and Wang, 1967), xiii, 12. Nell Irvin Painter, *Standing at Armageddon: The United States, 1877-1919* (New York: Norton, 1987). Maury Klein, *The Flowering of the Third America: The Making of an Organizational Society, 1850-1920* (Chicago: Ivan R. Dee, 1993). Joel Williamson, *A Rage for Order: Black-White Relations in the American South since Emancipation* (New York: Oxford University Press, 1986).

7. Mintz and Kellog, *Domestic Revolutions*, 109–110. Coontz, *Social Origins of Private Life*, 294. Seward, *American Family*, 92. Nancy Schrom Dye and Daniel Blake Smith, "Mother Love and Infant Death, 1750–1920," *Journal of American History* 73, no. 2 (September 1986): 329–353. Clifford, "Home and School in Nineteenth Century America," 7.
8. Daniel Scott Smith, "'Early' Fertility Decline in America: A Problem in Family History," *Journal of Family History* 12 (1987): 73–84. Peter Filene, *Him/her/self: Gender Identities in Modern America* (Baltimore: Johns Hopkins, 1998), 17. Paula Fass, *The Damned and the Beautiful: American Youth in the 1920s* (New York: Oxford, 1977), 57–80. Stevenson, *The Victorian Homefront*, 41.
9. Coontz, *Social Origins of Private Life*, 295–296, 150, 353–354, 349. Ruth Schwartz Cowan, *More Work for Mother: The Ironies of Household Technology from the Open Hearth to the Microwave* (New York: Basic Books, 1983). M. E. Moore, *Parent, Teacher, and School* (New York: MacMillan, 1923), 31. Glenna Matthews, *Just a Housewife: The Rise and Fall of Domesticity in America* (New York: Oxford, 1987), 92–115. Margaret Marsh, *Suburban Lives* (New Brunswick: Rutgers University Press, 1990).
10. Stevenson, *The Victorian Homefront*, xxv–xxvi. Gwendolyn Wright, *Building the Dream: A Social History of Housing in America* (New York: Pantheon, 1984).
11. Stevenson, *The Victorian Homefront*, 1–29.
12. Stevenson, *The Victorian Homefront*, 30–31, 41–43. Carl F. Kaestle, *Literacy in the United States: Readers and Reading Since 1880* (New Haven, Yale University Press, 1991), 280–282.
13. Ellen M. Plante, *Women at Home in Victorian America: A Social History* (New York: Facts on File, 1997), 100. Matthews, *Just a Housewife*, 177–178. Stevenson, *The Victorian Homefront*, 28–29.
14. Malcolm Cowley, *Exile's Return: A Literary Odyssey of the 1920's* (New York: Viking, 1951), 65 [tense altered]. Matthews, *Just a Housewife*, 86.
15. Colleen McDannell, *The Christian Home in Victorian America, 1840–1900* (Bloomington: Indiana University Press, 1986), 52–77. John Bodnar, *The Transplanted: A History of Immigrants in Urban America* (Bloomington: Indiana University Press, 1985). Cowley, *Exile's Return*, 64.
16. Fass, *The Damned and the Beautiful*, 42. T. J. Jackson Lears, *No Place of Grace: Antimodernism and the Transformation of American Culture, 1880–1920* (New York: Pantheon, 1981), 48.
17. Peter N. Stearns, *Anxious Parents: A History of Modern Childrearing in America* (New York: New York University Press, 2003), 176.
18. Mintz and Kellog, *Domestic Revolutions*, 96. Allan Carlson, *The New Agrarian Mind: The Movement toward Decentralist Thought in Twentieth-Century America* (New Brunswick: Transaction, 2000), 2–3.

19. Jean Gordon and Jan McArthur, "American Women and Domestic Consumption, 1800–1920: Four Interpretive Themes" in Marilyn Ferris Motz and Pat Browne, eds., *Making the American Home: Middle Class women and Domestic Material Culture, 1840–1940* (Bowling Green: Bowling Green State University Press, 1988), 34. Lears, *No Place of Grace*, 28, xiii. Daniel E. Sutherland, *The Expansion of Everyday Life, 1860–1876* (New York: Harper and Row, 1989), 46.
20. Filene, *Him/her/self*, 46. George Marsden, *Fundamentalism and American Culture* (New York: Oxford University Press, 2006), 240–241. Fraser, *Between Church and State*, 116–126. After Sputnik, the 1957 rocket symbolizing Russian preeminence in the space-race, the Federal Government became more serious about science education and created the Biological Sciences Curriculum Study series (BSCS), a rigorous curriculum written by top-tier scientists for American high schools. For the first time in decades, evolution was back in the nation's schools. The creation science movement was born as a reaction against this development in the mid 1960s. See Edward J. Larson, *Summer for the Gods: The Scopes Trial and America's Continuing Debate Over Science and Religion* (New York: Basic Books, 1997), 248–250.
21. Filene, *Him/her/self*, 44. Steven Selden, *Inheriting Shame: The Story of Eugenics and Racism in America* (New York: Teachers College Press, 1999), 22–33.
22. Ashby, *Endangered Children*, 94–96. Coontz, *Social Origins of Private Life*, 263. Mintz, "Regulating the American Family," 396–397. Stephen Provasnik, "Judicial Activism and the Origins of Parental Choice: The Court's Role in the Institutionalization of Compulsory Education in the United States, 1891–1925," *History of Education Quarterly* 46, no. 3 (Fall 2006): 311–347. Allan Carlson, *The "American Way": Family and Community in the Shaping of the American Identity* (Wilmington: ISI Books, 2003), 48–54.
23. Carlson, *The "American Way,"* 64–75. Morton Keller, *Affairs of State: Public Life in Late Nineteenth Century America* (Cambridge: Belknap, 1977), 128. Mintz and Kellog, *Domestic Revolutions*, 127. Mintz, "Regulating the American Family," 395.
24. Gordon and McArthur, "American Women and Domestic Consumption," 38–39. Fass, *The Damned and the Beautiful*, 101. Beatty, *Preschool Education in America*, 151. Julia Grant, *Raising Baby by the Book: The Education of American Mothers* (New Haven: Yale University Press, 1998), 3–4.
25. William W. Cutler, *Parents and Schools: The 150-Year Struggle for Control in American Education* (Chicago: University of Chicago Press, 2000), 8–9. Megan Birk, "Playing House: Training Modern Mothers at Iowa State College Home Management Houses, 1925–1958," *The Annals of Iowa* 64, no. 1 (Winter 2005): 37–66. Dorothy W. Baruch, "This 'Parental Instinct,'" *National Parent-Teacher* 29, no. 6 (February 1935): 9.
26. Beatty, *Preschool Education in America*, 138–156, 48–50, 93.

27. Lawrence Cremin, *American Education: The Metropolitan Experience, 1876–1980* (New York: Harper and Row, 1988), 546. Claudia Goldin and Lawrence F. Katz, "Human Capital and Social Capital: The Rise of Secondary Schooling in America, 1910–1940," *Journal of Interdisciplinary History* 29, no. 4 (Spring 1999): 685. Fass, *The Damned and the Beautiful*, 55–56. David Tyack, *The One Best System: A History of American Urban Education* (Cambridge: Harvard University Press, 1974). Tyack and Hansot, *Learning Together*, 165–200.
28. Cremin, *American Education: The Metropolitan Experience*, 551. Paula Fass, *Outside In: Minorities and the Transformation of American Education* (New York: Oxford University Press, 1989), 13–35. Ira Katznelson and Margaret Weir, *Schooling for All: Class, Race, and the Decline of the Democratic Ideal* (Berkeley: University of California Press, 1988). Fass, *The Damned and the Beautiful*, 6.
29. Stephen Lassonde, *Learning to Forget: Schooling and Family Life in New Haven's Working Class, 1870–1940* (New Haven: Yale University Press, 2005). David P. Baker, "Schooling All the Masses: Reconsidering the Origins of American Schooling in the Postbellum Era," *Sociology of Education* 72, no. 4 (October 1999): 209.
30. Goldin and Katz, "Human Capital and Social Capital," 683–723. Frank, *Life with Father*, 140.
31. Cutler, *Parents and Schools*, 17–18. Paul C. Gutjahr, *An American Bible: A History of the Good Book in the United States, 1777–1880* (Stanford: Stanford University Press, 1999), 113–143. Brian Gill and Steven Schlossman, "A Sin Against Childhood: Progressive Education and the Crusade to Abolish Homework, 897–1941," *American Journal of Education* 105 (November 1996): 29–38.
32. Cutler, *Parents and Schools*, 27, 24. Mary Harmon Weeks, ed., *Parents and their Problems: Child Welfare in Home, School, Church, and State* (Washington, D.C.: National Congress of Mothers and Parent-Teacher Associations, 1914), 106. Larry Cuban's *How Teachers Taught: Constancy and Change in American Classrooms, 1890–1980* (New York: Longman, 1984), for example, ignores the parental factor in his otherwise compelling explanation of why so many educational reforms failed.
33. John W. Whitehead and Alexis Irene Crow, *Home Education: Rights and Reasons* (Wheaton, Ill: Crossway, 1993), 121–123. Sharon Nalbone Richardson and Perry A. Zirkel, "Home Schooling Law" in Jane Van Galen and Mary Anne Pitman, eds., *Home Schooling: Political, Historical, and Pedagogical Perspectives* (Norwood, NJ: Ablex Publishing Corporation, 1991), 174–177. James W. Tobak and Perry A. Zirkel, "Home Instruction: An Analysis of the Statutes and Case Law," *University of Dayton Law Review* 8, no. 1 (Fall 1982), 23–24. Michael LaMorte, *School Law: Cases and Concepts* (Boston: Allyn and Bacon, 1999), 20–22.

34. Lawrence Kotin and William F. Aikman, *Legal Foundations of Compulsory School Attendance* (Port Washington, NY: Kennikat Press, 1980), 85.
35. Coontz, *Social Origins of Private Life*, 273. Hiram Orcutt, *The Parents' Manual: Or, Home and School Training* (Boston: Thompson, 1874), 90–91. John Dewey, *Democracy and Education* (New York: MacMillan, 1916), 18. David Reisman, *The Lonely Crowd: A Study of the Changing American Character* (New Haven: Yale University Press, 1950), 9, 14–17. Lears, *No Place of Grace*, 17. Colleen McDannell, "Parlor Piety: The Home as Sacred Space in Protestant America" in Jessica Foy and Thomas J. Schlereth, eds., *American Home Life, 1880–1930* (Knoxville: University of Tennessee Press, 1992), 162–189.
36. Diana Korzenik, *Drawn to Art: A Nineteenth-Century American Dream* (Hanover: University of New England Press, 1985). Eustace Broom and Bertha Trowbridge, "The Visiting Teacher's Job," *The Elementary School Journal* 26, no. 9 (May 1926), 656, 657.
37. Kaestle, *Literacy in the United States*, 52–63, 229–230, 241.
38. Kaestle, *Literacy in the United States*, 236–237, 243. Barbara Sicherman, "Reading and Ambition: M. Carey Thomas and Female Heroism," *American Quarterly* 45, no. 1 (March 1993): 73–103.
39. Patricia A. Lynott, "The Education of the Thirteenth Apostle: Susan Young Gates, 1856–1933," *Vitae Scholasticae* 16 (Fall 1997): 72–90.
40. John B. Judis, *William F. Buckley, Jr.: Patron Saint of the Conservatives* (New York: Simon and Schuster, 1988), 19–32. William F. Buckley, Jr., *Nearer, My God: An Autobiography of Faith* (New York: Doubleday, 1997), 2.
41. West, *Growing Up with the Country*, 188, 184. Theodore Prey Jorgensen, "An Autobiography: The Adventures of a Physicist" (Unpublished Manuscript, 1996, in possession of author), 7–8.
42. West, *Growing Up with the Country*, 183, 180–181. Joanna L. Stratton, *Pioneer Women: Voices from the Kansas Frontier* (New York: Simon and Schuster, 1981), 157–160. Sutherland, *The Expansion of Everyday Life*, 75.
43. Sutherland, *The Expansion of Everyday Life*, 54–55. West, *Growing Up with the Country*, 73–98, 58, 62–64.
44. West, *Growing Up with the Country*, 189–192. "A Century of Tradition and Innovation," *Calvert School: Homeschooling.* <http://www.calvertschool.org/engine/content.do?BT_CODE=CES1534> (September 26, 2006). Kate Tsubata, "Learning From Calvert School," *Washington Times* (March 20, 2006): sec. B.
45. "Teaching Children at Home," *Time Magazine* 112 (December 4, 1978): 78. "History," *Griggs University and International Academy*, <http://www.hsi.edu/history.html> (May 25, 2007). Alayne Thorpe, telephone conversation with author (July 18, 2007).
46. Harriet F. Bergmann, "'The Silent University': The Society to Encourage Studies at Home, 1873–1897," *New England Quarterly* 74, no. 3 (September 2001): 447–477.

47. James D. Watkinson, "Education for Success: The International Correspondence Schools of Scranton, Pennsylvania," *The Pennsylvania Magazine of History and Biography* 120, no. 4 (October 1996): 343–369.
48. Robert L. Hampel, "Men of Business: The National Home Study Council, 1928–1941" (paper presented at the annual meeting of the History of Education Society, Ottawa, Canada, October 27, 2006).
49. Cremin, *American Education: The Metropolitan Experience*, 77–80. Ruth Hutchinson Crocker, *Social Work and Social Order: The Settlement Movement in Two Industrial Cities, 1889–1930* (Urbana: University of Illinois Press, 1992). Elisabeth Lasch-Quinn, *Black Neighbors: Race and the Limits of Reform in the American Settlement House Movement, 1890–1945* (Chapel Hill: University of North Carolina Press, 1993).
50. Ashby, *Endangered Children*, 35–54, 102. Priscilla Ferguson Clement, "Families and Foster Care: Philadelphia in the Late Nineteenth Century" in N. Ray Hiner and Joseph M. Hawes, eds., *Growing Up in America: Children in Historical Perspective* (Urbana: University of Illinois Press, 1985), 137, 139.
51. Judith Sealander, *The Failed Century of the Child: Governing America's Young in the Twentieth Century* (New York: Cambridge University Press, 2003), 359–360. Valerie Quinney, "Childhood in a Southern Mill Village," *International Journal of Oral History* 3, no. 3 (November 1982), 183, 171–179. Eunice G. Pollack, "The Childhood We Have Lost: When Siblings were Caregivers, 1900–1970," *Journal of Social History* 36, no. 1 (Fall 2002): 31–61.

Chapter Four   Why Homeschooling Happened, 1945–1990

1. Marion Schickel, telephone interview with author (November 6, 2006).
2. Sealander, *The Failed Century of the Child*, 187. Hine, *Rise and Fall of the American Teenager*, 214. Lawrence Cremin, *American Education: The Metropolitan Experience*, 544–551. Kaestle, *Literacy in the United States*, 283.
3. Kenneth T. Jackson, *Crabgrass Frontier: The Suburbanization of the United States* (New York: Oxford, 1985), 173–175.
4. Jackson, *Crabgrass Frontier*, 162–168, 233, 4. Dolores Hayden, *Building Suburbia: Green Fields and Urban Growth, 1820–2000* (New York: Pantheon, 2003), 165–167.
5. Jackson, *Crabgrass Frontier*, 233, 293–295. Hayden, *Building Suburbia*, 163–164.
6. Hayden, *Building Suburbia*, 148–151. Elaine Tyler May, *Homeward Bound: American Families in the Cold War Era* (New York: Basic Books, 1988), 169–171. Beth Bailey and David Farber, *The Columbia*

Guide to America in the 1960s (New York: Columbia University Press, 2003), 5–6. Jackson, *Crabgrass Frontier*, 279–281.
7. Jim Cullen, *The American Dream: A Short History of an Idea That Shaped a Nation* (New York: Oxford, 2003), 152. Jackson, *Crabgrass Frontier*, 241–242, 289. Hayden, *Building Suburbia*, 165–167. Richard J. Altenbaugh, "Liberation and Frustration: Fifty Years after *Brown*," *History of Education Quarterly* 44, no. 1 (Spring 2004): 3–7. Andrew Weise, *Places of Their Own: African-American Suburbanization in the Twentieth Century* (Chicago: University of Chicago Press, 2004), 1, 222–229.
8. May, *Homeward Bound*, 181. Filene, *Him/her/self*, 198. Grant, *Raising Baby by the Book*, 203. Joanne Meyerowitz, "Women, Cheesecake, and Borderline Material: Responses to Girlie Pictures in the Mid-Twentieth Century U.S.," *Journal of Women's History* 8, no. 3 (Fall 1996): 9–36. Susan Lynn, "Gender and Progressive Politics: A Bridge to Social Activism of the 1960s" in Joanne Meyerowitz, ed., *Not June Cleaver: Women and Gender in Postwar America, 1945–1960* (Philadelphia: Temple University Press, 1994), 103–127. Michelle Nickerson, "'The Power of a Morally Indignant Woman': Republican Women and the Making of California Conservatism," *Journal of the West* 42, no. 3 (July 2003): 35–43. Sylvie Murray, *The Progressive Housewife: Community Activism in Suburban Queens, 1945–1965* (Philadelphia: University of Pennsylvania Press, 2003), 12.
9. Ella Taylor, *Prime-Time Families: Television Culture in Postwar America* (Berkeley: University of California Press, 1989). Joanne Meyerowitz, "Beyond the Feminine Mystique: A Reassessment of Postwar Mass Culture, 1946–1958" in Meyerowitz, ed., *Not June Cleaver* (Philadelphia: Temple University Press, 1994), 229–262. Julia Kirk Blackwelder, *Now Hiring: The Feminization of Work in the United States, 1900–1995* (College Station: Texas A&M University Press, 1997). Beth Bailey, "She 'Can Bring Home the Bacon': Negotiating Gender in Seventies America" in Beth Bailey and David Farber, eds., *America in the Seventies* (Lawrence: University Press of Kansas, 2004), 108. Filene, *Him/her/self*, 177–181. Sara M. Evans, *Born for Liberty: A History of Women in America* (New York: Free Press, 1989), 253–254.
10. Mintz and Kellogg, *Domestic Revolutions*, 186. Filene, *Him/her/self*, 187. David Steigerwald, *The Sixties and the End of Modern America* (New York: St. Martin's, 1995). May, *Homeward Bound*, 136–161.
11. Grant, *Raising Baby by the Book*, 204–205, 218–227, 244. Nancy Pottishman Weiss, "Mother, the Invention of Necessity: Dr Benjamin Spock's *Baby and Child Care*" in Hiner and Hawes, eds., *Growing Up in America* (Urbana: University of Illinois Press, 1985), 283–303. Virginia Rogers, "Should Parents Help with Homework?" *National Parent Teacher* 46, no. 3 (November 1951): 4. Esther Prevy, "NPT Quiz Program," *National Parent Teacher*, 43, no. 1 (September 1948): 25.

12. O. Spurgeon English, "Troubled Parent, Troubled Child," *National Parent Teacher* 45, no. 8 (April 1951): 4. Ann Hulbert, *Raising America: Experts, Parents, and a Century of Advice about Children* (New York: Knopf, 2003), 191–224. Thomas Frank, *The Conquest of Cool: Business Culture, Counterculture, and the Rise of Hip Consumerism* (Chicago: University of Chicago Press, 1997), 11–13.
13. Filene, *Him/her/self*, 206. Mintz and Kellogg, *Domestic Revolutions*, 188.
14. Hulbert, *Raising America*, 258, 328, 335–337. Weiss, "Mother, the Invention of Necessity," 302.
15. Hulbert, *Raising America*, 334, 293–324. Jeane Westin, *The Coming Parent Revolution* (Chicago: Rand McNally, 1981), 40. Bruce J. Schulman, *The Seventies: The Great Shift in American Culture, Society, and Politics* (New York: Free Press, 2001), 187–188. Daniel Scott Smith, "Recent Change and the Periodization of American Family History," *Journal of Family History* 20, no. 4 (1995): 338–340.
16. Filene, *Him/her/self*, 228. Steigerwald, *The Sixties and the End of Modern America*, 256. Hulbert, *Raising America*, 296.
17. David Farber, "The Torch had Fallen" in Beth Bailey and David Farber, eds., *American in the Seventies* (Lawrence: University Press of Kansas, 2004), 10. J. William Rioux and Stuart A. Sandow, *Children, Parents, and School Records* (Columbia, MD: National Committee for Citizens in Education, 1974). Lawrence Cremin, *Popular Education and its Discontents* (New York: Harper and Row, 1989), 5–6. B. Edward McClellan, *Moral Education in America: Schools and the Shaping of Character from Colonial Times to the Present* (New York: Teachers College Press, 1999), 76–77.
18. Sealander, *The Failed Century of the Child*, 356. Bailey, "She 'Can Bring Home the Bacon,'" 108. Philip Jenkins, *Decade of Nightmares: The End of the Sixties and the Making of Eighties America* (New York: Oxford University Press, 2006), 14, 111–133. Barbara Finkelstein, "Uncle Sam and the Children: A History of Government Involvement in Child Rearing" in Hiner and Hawes, eds., *Growing Up in America*, 255–266.
19. Not all scholars have agreed that the United States is divided. For a contrarian view see Morris P. Fiorina, *Culture War? The Myth of a Polarized America* (New York: Longman, 2005). Schulman, *The Seventies*, 246–252, 77, 16–17. David Farber, "Democratic Subjects in the American Sixties: National Politics, Cultural Authenticity, and Community Interest" in David Farber and Jeff Roche, eds., *The Conservative Sixties* (New York: Peter Lang, 2003), 8–9. Jenkins, *Decade of Nightmares*, 21. Steigerwald, *The Sixties and the End of Modern America*, 277.
20. Schulman, *The Seventies*, 12–14
21. Marianne DeKoven, *Utopia Limited: The Sixties and the Emergence of the Postmodern* (Durham: Duke University Press, 2004).
22. Schulman, *The Seventies*, 88–91.

23. Rosabeth Moss Kanter, "Communes" in Michael Gordon, eds, *The Nuclear Family in Crisis: The Search for an Alternative* (New York: Harper and Row, 1972), 174. Timothy Miller, *The Sixties Communes: Hippies and Beyond* (Syracuse: Syracuse University Press, 1999), 202–205. David E. Smith and James L. Sternfield, "Natural Child Birth and Cooperative Child Rearing in Psychedelic Communes" in Gordon, ed., *The Nuclear Family in Crisis: The Search for an Alternative* (New York: Harper and Row, 1972), 201.
24. Miller, *Sixties Communes*, 187–188. Bennett M. Berger, "Child Rearing in Communes" in Louise Kapp Howe, ed., *The Future of the Family* (New York: Simon and Schuster, 1972), 163.
25. Smith and Sternfield, "Natural Child Birth and Cooperative Child Rearing," 200–201. Berger, "Child Rearing in Communes," 169. John Rothchild and Susan Wolf, *Children of the Counterculture* (Garden City: Doubleday, 1976), 9, 191, quoted in Miller, *Sixties Communes*, 185.
26. Kenneth Cmiel, "The Politics of Civility" in David Farber, ed., *The Sixties: From Memory to History* (Chapel Hill: University of North Carolina Press, 1994), 271. Miller, *Sixties Communes*, 188–189. Eleanor Agnew, *Back from the Land: How Young Americans Went to Nature in the 1970s, and Why They Came Back* (Chicago: Ivan R. Dee, 2004), 43.
27. Miller, *Sixties Communes*, 239, 185.
28. Chelsea Cain, ed., *Wild Child: Girlhoods in the Counterculture* (Seattle: Seal Press, 1999), xvii, 56, 57, 99–100.
29. Cmiel, "Politics of Civility," 272–273. Agnew, *Back from the Land*, ix.
30. Frank, *Conquest of Cool*, 29–31. James Davison Hunter, *The Death of Character: Moral Education in an Age without Good or Evil* (New York: Basic Books, 2000), 81–148. Maurice Isserman and Michael Kazin, *America Divided: The Civil War of the 1960s* (New York: Oxford University Press, 2000), 247.
31. Tom Wolfe, "The Me Decade and the Third Great Awakening" in *The Purple Decades: A Reader* (New York: Farrar, Straus, Giroux, 1982), 265–296. Larry Norman, *Only Visiting This Planet* © 1972, Solid Rock Records.
32. Robert Wuthnow, *The Restructuring of American Religion: Society and Faith Since World War II* (Princeton: Princeton University Press, 1988), 132–172. Schulman, *The Seventies*, 92–95. Isserman and Kazin, *America Divided*, 241–245. Kaestle, *Literacy in the United States*, 287.
33. Scott Flipse, "Below-the-Belt Politics: Protestant Evangelicals, Abortion, and the Foundation of the New Religious Right, 1960–1975" in Farber and Roche, eds., *The Conservative Sixties*, 136–137. Farber, "The Torch Had Fallen," 23–25.
34. Evelyn A. Schlatter, "'Extremism in the Defense of Liberty': The Minutemen and the Radical Right" in Farber and Roche, eds., *The Conservative Sixties*, 37–50. Michelle Nickerson, "Moral Mothers and

Goldwater Girls" in Farber and Roche, eds., *The Conservative Sixties*, 53. Jonathan M. Schoenwald, "We Are an Action Group: The John Birch Society and the Conservative Movement in the 1960s" in Farber and Roche, eds., *The Conservative Sixties*, 21–36.
35. Schulman, *The Seventies*, 105–117. Schoenwald, "We Are an Action Group," 31–36.
36. Colleen McDannell, "Creating the Christian Home: Home Schooling in Contemporary America" in David Chidester and Edward T. Linenthal, eds., *American Sacred Space* (Bloomington: Indiana University Press, 2005), 187–219. Nickerson, "Moral Mothers and Goldwater Girls," 51–60.
37. Donald T. Critchlow, "Conservatism Reconsidered: Phyllis Schlafly and Grassroots Conservatism" in Farber and Roche, eds., *The Conservative Sixties*, 108–115.
38. Critchlow, "Conservatism Reconsidered," 116–118.
39. Critchlow, "Conservatism Reconsidered," 119–125. Donald Critchlow, *Phyllis Schlafly and Grassroots Conservatism, A Woman's Crusade* (Princeton: Princeton University Press, 2005).
40. Sara Diamond, *Not by Politics Alone: The Enduring Influence of the Christian Right* (New York: Guilford, 1998), 63–66, 131–155. Flipse, "Below-the-Belt Politics," 127–141.
41. Fraser, *Between Church and State*, 190–191.
42. Nickerson, "Moral Mothers and Goldwater Girls," 57. Joan DelFattore, *The Fourth R: Conflicts Over Religion in America's Public Schools* (New Haven: Yale University Press, 2004), 79.
43. Evans, cited in Michael W. Fuquay, "Civil Rights and the Private School Movement in Mississippi, 1964–1971," *History of Education Quarterly* 42, no. 2 (Summer 2002): 159–180, Evans is cited on p. 164.
44. Jason Bivins, *The Fracture of Good Order: Christian Antiliberalism and the Challenge to American Politics* (Chapel Hill: University of North Carolina Press, 2003), 81. James C. Carper, "The Christian Day School" in James C. Carper and Thomas C. Hunt, eds., *Religious Schooling in America* (Birmingham: Religious Education Press, 1984), 114–115. Susan D. Rose, *Keeping Them Out of the Hands of Satan: Evangelical Schooling in America* (New York: Routledge, 1988), 35–38. E. Vance Randall, "Religious Schools in America: Worldviews and Education" in Thomas C. Hunt and James C. Carper, eds., *Religion and Schooling in Contemporary America: Confronting Our Cultural Pluralism* (New York: Garland, 1997), 96–98. James C. Carper and Thomas C. Hunt, *The Dissenting Tradition in American Education* (New York: Peter Lang, 2007), 203–204.
45. *State v. Shaver*, 294 N.W.2d 883 (N.D. 1980). Kenneth Gangel, "How Did Christian School Education Begin?" in H. Wayne House, ed., *Schooling Choices* (Portland: Multnomah, 1988), 106, 75. Deborah Kovach Caldwell, "Conservative Christians Urge Parents to Help Bring Down Public School System," *Dallas Morning News*, February 3, 1999,

Newspaper Source Publications, EBSCOhost, (November 17, 2006). House, *Schooling Choices*, 19–167.
46. Filene, *Him/Her/self*, 223. Mark Hamilton Lytle, *America's Uncivil Wars: The Sixties Era from Elvis to the Fall of President Nixon* (New York: Oxford, 2006), 282. Michael S. Shepherd, "Home Schooling: A Legal View" in Anne Pedersen and Peggy O'Mara, eds., *Schooling at Home: Parents, Kids, and Learning* (Santa Fe: John Muir, 1990): 57.
47. "The Judicial Role in Attacking Racial Discrimination in Tax-Exempt Private Schools," *Harvard Law Review* 93, no. 2 (December 1979): 378–384. William Martin, *With God on Our Side: The Rise of the Religious Right in America* (New York: Broadway Books, 1996), 168–173.
48. Linda Dobson, *Homeschoolers' Success Stories*, 6. Dobson's authority for the claim was Mark Hegener, who arrived at this conclusion as a result of hours of phone conversations with support group leaders around the country who didn't know what to do with the massive influx of fundamentalist Christians into their midst in the early 1980s. He later admitted that the tax code issue had been an inference on his part. Mark Hegener, telephone conversation with author (July 12, 2007). James C. Carper and Jack Layman, "Independent Christian Day Schools: The Maturing of a Movement," *Catholic Education* 5, no. 4 (June 2002): 505. Carper and Hunt, *The Dissenting Tradition in American Education*, 233.
49. Schulman, *The Seventies*, xvi. Steigerwald, *The Sixties and the End of Modern America*, 244.
50. Mitchell Stevens, *Kingdom of Children: Culture and Controversy in the Homeschooling Movement* (Princeton: Princeton University Press, 2001), 8. The Bill Gaither Trio, "I am a Promise," *Especially for Children: Classic Moments from the Bill Gaither Trio*, Benson Records (1994). Jenkins, *Decade of Nightmares*, 134–151.
51. James C. Carper, "Pluralism to Establishment to Dissent: The Religious and Educational Context of Home Schooling," *Peabody Journal of Education* 75 (2000): 8–19.

Chapter Five   Three Homeschooling Pioneers

1. David Fleisher and David M. Fredman, *Death of an American: The Killing of John Singer* (New York: Continuum, 1983). Eugene Linden, "The Return of the Patriarch!" *Time* (February 1, 1988): 21. Guy Murray, "From the Utah History Archives," *Messenger and Advocate* (September 26, 2006). <http:// http://messengerandadvocate.wordpress.com/2006/09/26/from-the-utah-history-archives/> (February 16, 2007).
2. On the Perchemlides case, see Richard A. Bumstead, "A Landmark Decision for Home Education" in Theodore E. Wade, Jr., Dorothy N. Moore, and Richard A. Bumstead, eds., *School at Home: How Parents Can Teach Their Own Children* (Colfax, CA: Gazelle Publications,

1980), 7–15; Stephen Arons, "Is Educating Your Own Child a Crime?" *Saturday Review* 5 (November 25, 1978): 16–20; Stephen Arons, "Public School Meltdown," *Michigan Law Review* 79, no. 4 (March 1981): 792–801.
3. Allan Carlson, "Household Freedom and Home Education: New Agrarian Dreams for the Twenty-First Century," *The Cresset: A Review of Literature, the Arts, and Public Affairs* (Lent, 2003): 5–10. Ralph Borsodi, *Flight from the City: The Story of a New Way to Family Security* (New York: Harper and Brothers, 1933). Rita Scherman, *A Mother's Letters to a Schoolmaster* (New York: A. A. Knopf, 1923). William B. Barrett, *The Home Education of a Boy: Being the True Account of a Father's Experiment in the Education of His Son in His Home* (Scarsdale: Updegraff Press, 1950). Phil Carden, "Running a Home in a Schoolhouse: Something New in Tennessee Education," *The Nashville Tennessean Magazine* (June 7, 1953) <http://www.tcarden.com/calvert/calvert.cfm> (July 20, 2005).
4. Hal Bennett, *No More Public School* (New York: Random House, 1972). Howard Rowland, *No More School* (New York: Dutton, 1975). Alvin Toffler, *Future Shock* (New York: Random House, 1970). Rosa Covington Packard, *The Hidden Hinge* (Notre Dame: Fides, 1972).
5. Bivins, *Fracture of Good Order*, 94–96.
6. Peter Schrag and Diane Divocky, *The Myth of the Hyperactive Child: And Other Means of Child Control* (New York: Pantheon, 1975). Hine, *Rise and Fall of the American Teenager*, 255. John E. Coons and Stephen D. Sugarman, *Education by Choice: The Case for Family Control* (Berkeley, UCA Press, 1978). Christopher Jencks, "Education Vouchers," *The New Republic* 163 (July 4, 1970): 19–21.
7. Casey Patrick Cochran, "The Home Schooling Movement in the U.S.: Georgia as a Test Case" (Ph.D. diss., Emory University, 1995), 27–28.
8. Cochran, "Home Schooling Movement in the U.S.," 29–30.
9. Pat Farenga, "Homeschooling and John Holt's Vision" in Mary M. Leue, ed., *Challenging the Giant: The Best of SKOLE* (Albany, NY: Down-to-Earth Books, 1992), 198–199. Ron Miller, *Free Schools, Free People* (Albany: State University of New York Press, 2002), 80–81. Peter Marin, review of *Freedom and Beyond*, by John Holt, *Learning* 1, no. 1 (November 1972): 90.
10. Mel Allen, "The Education of John Holt," *Yankee* (December 1981): 150–153.
11. Miller, *Free Schools*, 81. "Photographs of John Holt," *HoltGWS.com, John Holt and Growing without Schooling*, <http://www.holtgws.com/jhphotosp.2.html> (April 6, 2007). Diane Divoky, "John Holt: Where are you Now?" *Learning* 4, no. 3 (November 1975): 11–12.
12. Stevens, *Kingdom of Children*, 24–25. Farenga, "Homeschooling and John Holt's Vision," 208–209. It should be noted that for Holt, unschooling simply meant not going to school. Only gradually did the term come to designate only the portion of homeschoolers who rejected formal

curricula. See Aaron Falbel, "Uncomfortable with the H-Word," *Growing without Schooling* 118 (September/October 1997): 33–34.
13. Mel Allen, "The Education of John Holt," 153–155.
14. The *New York Times Magazine* had run an earlier piece by Patricia Heidenry titled "Home Is Where the School Is" on October 19, 1975. *Time*'s piece was followed by similar articles in *Newsweek* in 1979, and *Parents* and *U.S. News* in 1980. Prior to these stories nearly all articles on education in the home appearing in popular magazines were about the pros and cons of homework. By the mid-1980s homeschooling had eclipsed homework as the most popular theme of articles in trade magazines about education in the home.
15. Farenga, *Growing without Schooling: A Record of A Grassroots Movement* (Cambridge, MA: Holt Associates, 1999), 127, 151. Helen Hegener, "Interview with Linda Dobson," *Home Education* 14, no. 5 (September–October 1997): 24–27. Darla Isackson, "Joyce Kinmont, Homeschooling Pioneer," *Meridian: The Place Where Latter-Day Saints Gather* (2005), <http://www.meridianmagazine.com/ideas/051006homeschool14.html> (May 24, 2007).
16. "Mixed Allies" in Farenga, *Growing without Schooling*, 25.
17. "Useful Resources" in Farenga, *Growing without Schooling*, 5.
18. Ellen G. White, *The Adventist Home: Counsels to Seventh-Day Adventist Families as Set Forth in the Writings of Ellen G. White* (Nashville: Southern Pub. Assn., 1952), 183, 185, 188, 194, 202–203.
19. *People of the State of Illinois v. Marjorie Levisen, et al.*, 404 Ill. 574, 14 A.L.R.2d 1364 (1950).
20. Cochran, "Home Schooling Movement in the U.S.," 55–59. Raymond S. Moore, "Dorothy…A Modern Dorcas," *Moore Report International* 14, no. 2 (March/April 2002): 1, 7.
21. Pat Gee, "Home-Schooling Trailblazer Dies, 86," *Honolulu Star-Bulletin* (March 1, 2002), <http://starbulletin.com/2002/03/01/news/story16.html> (May 28, 2007). "Timeline of the Life of Dorothy Lucile Nelson Moore," *Moore Report International* 14, no. 2 (March/April 2002): 7.
22. Casey Patrick Cochran, "From Early Childhood Education Critic to Home School Champion: The Curious Ascendance of Raymond Moore, 1961–1985" (Paper presented at History of Education Society Annual Meeting, Baltimore, October 2005).
23. Whitehead and Crow, *Home Education: Rights and Reasons*, 117–119. Raymond S. Moore and Dennis R. Moore, "The Dangers of Early Schooling," *Harper's* 245, no. 1466 (July 1972): 58–62. Raymond S. Moore and Dennis R. Moore, "When Should Your Child Go to School?" *Reader's Digest* 101, no. 606 (October 1972): 143–147. Raymond S. Moore and Dennis R. Moore, "Early Schooling for All?" *Congressional Record* 118, no. 167 (October 16, 1972): E8726–E8741.
24. Moore and Moore, "Dangers of Early Schooling," 62.

25. Cochran, "Home Schooling Movement in the U.S.," 60. "Letters," *Harpers* 245, no. 146 (September 1972): 3–5.
26. Buff Bradley, review of *Better Late Than Early*, by Raymond Moore and Dorothy Moore, *Learning* 4, no. 4 (December 1975): 61. Whitehead and Crow, *Home Education: Rights and Reasons*, 118. Cochran, "From Early Childhood Education Critic to Home School Champion."
27. James Dobson, "Foreword" in Raymond Moore and Dorothy Moore, *Home Grown Kids: A Practical Handbook for Teaching Your Children at Home* (Waco: Word, 1981), 10.
28. Gena Suarez, "A Tribute to a True Pioneer: Dr. Raymond Moore," *The Old Schoolhouse* (Spring 2003): 2–3. Raymond Moore and Dorothy Moore, *Home Grown Kids: A Practical Handbook for Teaching Your Children at Home* (Waco: Word, 1981), 13–14. Ellen G. White, *The Adventist Home*.
29. Cochran, "From Early Childhood Education Critic to Homeschool Champion."
30. John Holt, "How Many Are We?" *Growing without Schooling* 32 (April 1, 1983): 7–8.
31. Mark R. Rushdoony, "The Vision of R. J. Rushdoony," Chalcedon Foundation (October 17, 2005), <http:// www.chalcedon.edu/articles/article.php?ArticleID=185> (February 28, 2007).
32. Jeff Sharlet, "Through a Glass Darkly: How the Christian Right is Reimagining U.S. History," *Harpers* 313, no. 1879 (December 2006): 33–43.
33. Marsden, *Fundamentalism and American Culture* (2006), 246–248, 328.
34. "Extreme Makeover: Calif. Theocrat Seeks Kinder, Gentler Image," *Church and State* 57, no. 9 (Oct 2004), 20–21. Ross Douthat, "Theocracy, Theocracy, Theocracy," *First Things* 165 (August/September 2006): 24.
35. Michael Farris, *The Joshua Generation: Restoring the Heritage of Christian Leadership* (Nashville: Broadman and Holman, 2005), 11.
36. Rousas John Rushdoony, *The Messianic Character of American Education* (Nutley, NJ: Craig Press, 1963), 332.
37. Rousas John Rushdoony, *The Nature of the American System* (Vallecito, CA: Ross House Books, 1965), 26. Rousas John Rushdoony, *Institutes of Biblical Law*, vol. 1 (Vallecito, CA: Ross House Books, 1973), 185.
38. Rousas John Rushdoony, *The Philosophy of the Christian Curriculum* (Vallecito, CA: Ross House Books, 1981), 30, 29, 86.
39. Brian M. Abshire, "A Reconstructed View of Evangelism," *Chalcedon Report* 370 (May 1996): 22, 21. "Christian Home Schooling: Raising A Victorious Army for Jesus Christ," *Chalcedon Report* 392 (March 1998): front cover.

Chapter Six   The Changing of the Guard, 1983–1998

1. Jane Ann Van Galen, "Schooling in Private: A Study of Home Education" (Ph.D. diss., University of North Carolina, Chapel Hill, 1986). Tom Lauricella, "The Education of a Home Schooler," *Smartmoney* (November 2001): 119. Vernon L. Bates, "Lobbying for the Lord: The New Christian Right Home-Schooling Movement and Grassroots Lobbying," *Review of Religious Research* 33, no. 1 (September 1991): 7.
2. Patricia Lines, *Estimating the Home Schooled Population* (Washington, D.C.: Office of Educational Research and Improvement, 1991). Cochran, "Home Schooling Movement in the U.S.," 1–2. Mitchell Stevens, "Normalization of Homeschooling in the USA," *Evaluation and Research in Education* 17, no. 2, 3 (2003): 95. Scott W. Somerville, "The Politics of Survival: Home Schoolers and the Law," *Home School Legal Defense Association* (April 2001), <http://www.hslda.org/docs/nche/000010/politicsofsurvival.asp> (May 31, 2007). Isabel Lyman, *The Homeschooling Revolution* (Amherst: Bench Press International, 2000), 95.
3. Lyman, *The Homeschooling Revolution*, 96. Carlson, *New Agrarian Mind*, 211. Robert D. Putnam and Lewis M. Feldstein, *Better Together: Restoring the American Community* (New York: Simon and Schuster, 2003), 2–3. Mitchell Stevens, *Kingdom of Children*, 107–142.
4. Jane Van Galen, "Schooling in Private: A Study of Home Education." (Ph.D. diss., University of North Carolina, Chapel Hill, 1986). J. Gary Knowles, Stacey E. Marlow, and James A. Muchmore, "From Pedagogy to Ideology: Origins and Phases of Home Education in the United States, 1970–1990," *American Journal of Education* 100, no. 2 (February 1992): 195–235. Stevens, *Kingdom of Children*, 19–20.
5. Stevens makes this same point when he writes of inclusives, "by acceding to the ideal of inclusiveness, they also accede to a goal that excludes many believers." See *Kingdom of Children*, 136.
6. Quentin Johnson, "Christian Home Schooling: How it All Began," *Chalcedon Report* 392 (March 1998): 22–23. Frederick Clarkson, "No Longer without Sheep," *Political Research Associates* (March/June 1994), <http://www.publiceye.org/magazine/v08n1/chrisre3.html> (June 5, 2007).
7. Johnson, "Christian Home Schooling," 23. Farenga, *Growing without Schooling*, 181.
8. John Holt, "How Many Are We?" 8. Patricia M. Lines, "An Overview of Home Instruction," *Phi Delta Kappan* 68, no. 7 (March 1987): 512. Jon Davis, "Legacy of Church Leader Marked in Success of Home Schooling," *Arlington Heights Daily Herald* (May 26, 2002), <http://www.christianliberty.com/pastorPaulMemorial/pastorPaulMemorial/pastor

PaulMemorial9.html> (June 6, 2007). "Worldview: Philosophy of Education," CLASS Homeschools, <http://www.homeschools.org/worldview/philosophyOfChristianEducation.html#theGoalOfChristian Education> (June 6, 2007).
9. Gregg Harris, telephone conversation with author, August 14, 2007. Gregg Harris, *The Christian Home School* (Brentwood, TN: Wolgemuth and Hyatt, 1988), 54–55.
10. Harris, telephone conversation. Chris Klicka, *Home School Heroes: The Struggle and Triumph of Home Schooling in America* (Nashville: Broadman and Holman, 2006), xi, 21–22. Stevens, *Kingdom of Children*, 237.
11. Raymond Moore, "The Ravage of Home Education Through Exclusion by Religion," *A to Z Home's Cool Homeschooling* (October 1994), <http://homeschooling.gomilpitas.com/extras/WhitePaper.htm> (June 12, 2007).
12. Harris, *The Christian Home School*, 149.
13. Harris, *The Christian Home School*, 145, back cover. Gregg Harris, *The Christian Home School* (Gresham, OR: Noble Publishing Associates, 1995), back cover. Klicka, *Home School Heroes*, 21–22. Stevens, *Kingdom of Children*, 115–118.
14. "Bill Gothard," *Wikipedia* (February 19, 2007), <http://www.wikipedia.org/wiki/Bill_Gothard> (June 22, 2007). Bill Gothard, "Basic Life Principles: Seven Universal Principles of Life," *Institute in Basic Life Principles* <http:// iblp.org/iblp/seminars/basic/principles/> (June 26, 2007).
15. Seelhoff, "A Homeschooler's History, Part II," 68–70.
16. Rick Miesel, "Bill Gothard: General Teachings/Activities," *Biblical Discernment Ministries* (February 2004), <http://www.rapidnet.com/~jbeard/bdm/exposes/gothard/general.htm> (June 26, 2007).
17. Lines, "An Overview of Home Instruction," 513. Dan B. Fleming and Thomas C. Hunt, "The World as Seen by Students in Accelerated Christian Education Schools," *Phi Delta Kappan* 68, no. 7 (March 1987): 518–523. CLASS also used titles from Rod and Staff Publishers, a Mennonite outfit.
18. Daniel L. Turner, *Standing without Apology: The History of Bob Jones University* (Greenville S.C.: Bob Jones University Press, 1997): 283, 304, 437. John Holt, "Pensacola School" in *Growing without Schooling* 22 <http://www.unschooling.com/gws/?cat=20> (July 5, 2007). Lines, "Overview of Home Instruction," 512. Lyman, *The Homeschooling Revolution*, 67, 76.
19. Stevens, *Kingdom of Children*, 54. Lyman, *The Homeschooling Revolution*, 75–79.
20. Mary Pride, *The Way Home: Beyond Feminism, Back to Reality* (Westchester, IL: Crossway, 1985). Isabel Lyman, *Homeschooling: Back to the Future?* (Washington, D.C.: Cato Institute, 1998).

21. Cathy Duffy, "RE: Contact Us" (July 9, 2007), personal e-mail (July 9, 2007).
22. Pride, *The Way Home*, 221–222. David Brooks, "The New Red-Diaper Babies," *New York Times* (December 7, 2004): sec. Editorial. McDannell, "Creating the Christian Home," 210.
23. Stevens, *Kingdom of Children*, 119. Bates, "Lobbying for the Lord," 8.
24. Stevens, *Kingdom of Children*, 119–120.
25. J. Rousas Rushdoony, *Chalcedon Report* 405 (April 1999) quoted in Seelhoff, "A Homeschooler's History, Part I," 35. Ted Olsen, "The Dragon Slayer," *Christianity Today* 42, no. 14 (December 7, 1998): 36–42.
26. Olsen, "The Dragon Slayer," 36–42. Frederick Clarkson, *Eternal Hostility: The Struggle between Theocracy and Democracy* (Monroe, MN: Common Courage Press, 1997), 92–93.
27. Somerville, "The Politics of Survival." It should be noted that the Rutherford Institute has departed dramatically from its original reconstructionist vision. Whitehead, who gained national prominence as legal counsel for Paula Jones during her sexual harassment suit against President Bill Clinton, has in the past decade and a half moved farther and farther away from his earlier positions. The Rutherford Institute is now primarily a civil liberties organization, defending the rights of gays, Muslims, Buddhists, and any other underdog group, with as much zeal as it has always defended Christians.
28. Jason Vest, "Mike Farris: For God's Sake Does He Have a Prayer of Becoming Virginia's Lieutenant Governor? Yes—and Some Say That's the Problem," *Washington Post* (August 5, 1993): sec. C. "Patrick Henry College's Michael Farris," interview by Terri Gross, *Fresh Air*, National Public Radio (May 24, 2006). Bivins, *Fracture of Good Order*, 100. "History," Concerned Women for America, <http://www.cwfa.org/history.asp> (June 28, 2007). *Farris v. Munro*, 99 Wn.2d 326, 662 P .2d 821, (Wa. 1983). Stephen Bates, *Battleground: One Mother's Crusade, The Religious Right and the Struggle for Our Schools* (New York: Henry Holt, 1993), 114.
29. Klicka, *Home School Heroes*, 27–31. Stevens, *Kingdom of Children*, 3. Zan Tyler, "The Curtain Rises on HSLDA," *The Home School Court Report* 19, no. 1 (January/February 2003), <http://www.hslda.org/courtreport/V19N1/V19N103.asp> (July 10, 2007).
30. Josh Harris, "Q and A with Chris Klicka," *New Attitude* 1, no. 1 (Spring 1993), <http://www.ylcf.org/newattitude/1-1/> (June 28, 2007). Hugh Davis Graham, "The Storm over Grove City College: Civil Rights Regulation, Higher Education, and the Reagan Administration," *History of Education Quarterly* 38, no. 4 (Winter 1998): 407–429.
31. Walter Olson, "Invitation to a Stoning: Getting Cozy with the Theocrats," *Reason Magazine* (November 1998): 64, 62. Gary North, "The Intellectual Schizophrenia of the New Christian Right," *Christianity and Civilization* 1 (Spring, 1982): 25. Gary North, *The*

Sinai Strategy: Economics and the Ten Commandments (Tyler, TX: Institute for Christian Economics, 1986), 59–60. Harris, "Q and A with Chris Klicka."
32. Klicka, *Home School Heroes*, 10.
33. Klicka, *Home School Heroes*, 8–10. Harris, "Q and A with Chris Klicka."
34. Klicka, *Home School Heroes*, 10, 35. James Davison Hunter, *Culture Wars: The Struggle to Define America* (New York: Basic Books, 1991): 208. Seelhoff, "Homeschooler's History, Part I," 43. Stevens, *Kingdom of Children*, 178. HSLDA, phone conversation with author (July 2, 2007). It should be noted that HSLDA is quite casual about its membership data. The above figures may not be fully accurate. For examples of HSLDA's vacillating membership claims, see "Who Does HSLDA Represent?" *Home Schooling is Legal*, <http://hsislegal.com/represent.asp> (July 2, 2007).
35. Raymond Moore, "The Ravage of Home Education." Seelhoff, "Homeschooler's History, Part I," 42.
36. Matthew C. Moen, "From Revolution to Evolution: The Changing Nature of the Christian Right," *Sociology of Religion* 55, no. 3 (Autumn 1994): 345–357. Cheryl Lindsey Seelhoff, "A Homeschooler's History of Homeschooling, part III," *Gentle Spirit Magazine* 6, no. 11 (June/July 2000): 42. Klicka, *Home School Heroes*, 24–26. Cheryl Lindsey Seelhoff, "A Homeschooler's History of Homeschooling, Part VI," *Gentle Spirit Magazine* 7, no. 4 (February/March 2001): 1–6.
37. Stevens, *Kingdom of Children*, 118–120. Harris, *The Christian Home School*, 111–112.
38. Stevens, *Kingdom of Children*, 123–126, 23–29. "Leaders Serving Leaders," *The National Alliance of Christian Home Education Leadership*, <http://www.achel.org/login/index.cfm> (July 12, 2007). "The Alliance Speaker Referral System," *The National Alliance Speaker's Bureau*, <http://speakers.achel.org/login/index.cfm> (July 12, 2007).
39. Patrick Farenga, "Being Proactive, Not Reactive," *Growing without Schooling* 118 (September/October 1997): 34–35.
40. Patrick Farenga, "RE: GWS Question" (April 21, 2007), personal e-mail (April 21, 2007). "A Conversation with Pat Montgomery," *Home Education Magazine* 7, no. 5 (September/October 1990): 15–17, 49–52.
41. Seelhoff, "A Homeschooler's History, Part 1," 43. J. Richard Fugate, *Will Early Education Ruin Your Child?* (Temple, AZ: Alpha Omega, 1990).
42. Raymond Moore, *Home School Burnout: What it is, What Causes it, and How to Cure it* (Nashville: Wolgemuth and Hyatt, 1988). Harris, telephone conversation.
43. Karen Braun, "Getting to Know National Home Education Legal Defense," *The Old Schoolhouse Magazine* (Winter 2006), <http://

www.thehomeschoolmagazine.com/How_To_Homeschool/articles/getting_to_know_nheld.php> (July 27, 2007). Deborah Stevenson, telephone conversation with author (July 13, 2007).

44. Phil Kuntz, "Home-Schooling Movement Gives House a Lesson," *Congressional Quarterly* (February 26, 1994): 479. Klicka, *Home School Heroes*, 250–259. Somerville, "Legal Rights for Homeschooling Families," 139–143. Larry and Susan Kaseman, "Survey and Lobbyists Cause Problems for Homeschoolers," *Home Education Magazine* 14, no. 5 (September–October 1997): 14–19. Cheryl Lindsey Seelhoff, "A Homeschooler's History of Homeschooling, Part IV: H.R. 6," *Gentle Spirit Magazine* 7, no. 1 (August/September 2000): 54–60.
45. Steve Grove, "Reading, Writing, and Right-Wing Politics," *Boston Globe* (August 15, 2004): sec. D1. Stevens, *Kingdom of Children*, 176–177.
46. Sue Patterson, "Unity and Diversity among Homeschoolers," *National Home Education Network* (1999), <http://www.nhen.org/main/default.asp?id=191> (July 13, 2007). "NHEN-PR," *Yahoo! Groups*, <http://groups.yahoo.com/group/NHEN-PR/> (July 13, 2007).
47. Stevens, *Kingdom of Children*, 127, 85. See, for example, the extensive site *Homeschooling Is Legal*, <http://hsislegal.com/> (July 13, 2007).
48. Cheryl Lindsey Seelhoff, "Confronting the Religious Right," *Off Our Backs* 36, no. 3 (2006): 18–25. Cheryl Lindsey Seelhoff, "Some of My Story," *Women's Space* (February 10, 2002), <http://www.womensspace.org/WomynsStories/1.html> (July 16, 2007). Cheryl Lindsey Seelhoff, "A Homeschooler's History of Homeschooling, Part V," *Gentle Spirit Magazine* 7, no. 2 (October/November 2000): 1–6.
49. Seelhoff, "Homeschooler's History, Part V," 4. Shay Seaborne, "The Truth About Cheryl," *Home Education Magazine*, <http://www.homeedmag.com/seelhoffvs.welch/truth.html> (July 16, 2007).
50. Seaborne, "The Truth about Cheryl." Seelhoff, "Some of My Story."
51. Seaborne, "The Truth About Cheryl."
52. Helen Cordes, "Battle for the Heart and Soul of Home Schoolers," *Salon.com* (October 2, 2000), <http://archive.salon.com/mwt/feature/2000/10/02/homeschooling_battle/index4.html> (July 17, 2007). Gregg Harris, "About Me" (April 9, 2007), <http://greggharrisblog.blogspot.com/> (July 17, 2007). Brian D. Ray, *Worldwide Guide to Homeschooling* (Nashville: Broadman and Holman, 2003): 47. The judge, in *Seelhoff v. Welch*, repeatedly rejected efforts by the defense to claim biblical law as the legitimate grounds for adjudicating the case. See Seaborne, "The Truth About Cheryl."

Chapter Seven Making It Legal

1. Klicka, *Home School Heroes*, 11, 8. Brian C. Anderson, "An A for Home Schooling," *City Journal*, The Manhattan Institute (Summer 2000),

<http://www.city-journal.org/html/10_3_an_a_for_home.html> (July 17, 2007). Scott W. Somerville, "Legal Rights for Homeschool Families" in Bruce S. Cooper, ed., *Home Schooling in Full View* (Greenwich, CT: Information Age Publishing, 2005), 135. Shay Seaborne, "Does HSLDA Work With State Associations and Support the Autonomy of Homeschoolers?" *Home Schooling Is Legal*, <http://hsislegal.com/states.asp> (July 18, 2007). Raymond Moore, "The Ravage of Home Education."
2. Samuel L. Blumenfeld, a longtime reconstructionist critic of public education and homeschooling advocate, has argued that compulsory education itself is unconstitutional because it violates the thirteenth amendment's prohibition against involuntary servitude. See his "Are Compulsory School Attendance Laws Necessary? (Part 3)," *Freedom Daily* (May 1991), <http://www.fff.org/freedom/0591c.asp> (July 18, 2007).
3. Farenga, *Growing without Schooling*, 18–20. Klicka, *Home School Heroes*, 115.
4. Neil Devins, "State Regulation of Home Instruction: A Constitutional Perspective" in Thomas N. Jones and Darel P. Semler, eds., *School Law Update 1986* (Topeka: National Organization on Legal Problems of Education, 1986): 159–174. David Allen Peterson, "Home Education v. Compulsory Attendance Laws: Whose Kids Are They Anyway?" *Washburn Law Journal* 24, no. 2 (Winter 1985): 274–299. Perry A. Zirkel, "Constitutional Contours to Home Instruction: A Second View" in Thomas N. Jones and Darel P. Semler, eds., *School Law Update 1986* (Topeka: National Organization on Legal Problems of Education, 1986): 175–182. Richardson and Zirkel, "Home Schooling Law," 159–210. Alma C. Henderson, "The Home Schooling Movement: Parents Take Control of Educating their Children," *Annual Survey of American Law* 4 (1991): 985–1009. It has not been lost on religious conservatives that the 14th Amendment argument leads them into tricky waters. As Scott Somerville, formerly with HSLDA, has noted, "After all, the strongest precedent for a right to teach a child at home is *Roe v. Wade*. *Roe*'s right to abortion is explicitly grounded in a fundamental right to 'bear and raise children.'" If homeschoolers embrace a constitutional right to privacy for homeschooling, then it is harder to object when the same argument is used to support abortion rights or to strike down anti-sodomy statutes. See Somerville, "The Politics of Survival."
5. Albert N. Keim, ed., *Compulsory Education and the Amish: The Right Not To Be Modern* (Boston: Beacon, 1975). Cochran, "Home Schooling Movement in the U.S.," 4. Zirkel, "Constitutional Contours," 176–179. Richardson and Zirkel, "Home Schooling Law," 164–165. Russell Kirk, "Escaping the Teachers' Clutches," *National Review* 32 (September 5, 1980): 1086. Stephanie B. Goldberg, "Education Begins at Home," *ABA Journal* 79 (August 1993): 87. Klicka, *Home School Heroes*, 100–101. "The Good, the Bad, the Inspiring," *Marking the Milestones*, Home School Legal Defense Association, <http://www.hslda.org/about/history/good-bad-inspiring.asp> (July 19, 2007).

6. For an example of the constitutional claim, see Lyman, *The Homeschooling Revolution*, 33-34. Zirkel, "Constitutional Contours," 177. Perry A. Zirkel, "Home/School Cooperation?" *Phi Delta Kappan* 78, no. 9 (May 1997): 727-728. Henry C. T. Richmond III, "Home Instruction: An Alternative to Institutional Education," *Journal of Family Law* 18, no. 353 (1980): 378.
7. Cremin, *American Education: The Metropolitan Experience*, 297-298.
8. Tobak and Zirkel, "Home Instruction," 1-60. While Tobak and Zirkel offer the most comprehensive analysis, it should be noted that several other analysts in the early 1980s, while all using the typology of explicit, implied, and no statutory language for home instruction, nevertheless came up with slightly different tallies of states in each category, which was testimony to the vagueness of many of these laws at the time. See, for example, Brendan Stocklin-Enright, "The Constitutionality of Home Instruction: The Role of Parents, State, and Child," *Willamette Law Review* 18, no. 568 (1982): 609, and Richmond III, "Home Instruction: An Alternative to Institutional Education," 378-381.
9. Stocklin-Enright, "The Constitutionality of Home Instruction: The Role of Parents, State, and Child," 600-601. Arons, "Is Educating Your Own Child a Crime?" 18. Whitehead and Crow, *Home Education: Rights and Reasons*, 293-296.
10. "The State and the New Protestant Dissenters: The *Whisner* Decision" in Carper and Hunt, *The Dissenting Tradition in American Education*, 217-237.
11. Michael C. O'Laughlin, "An Investigation of Public School Leaders' Attitudes Toward Home schooling" (Ph.D. diss., University of Rochester, 1993). David Guterson, *Family Matters: Why Homeschooling Makes Sense* (New York: Harcourt, Brace, Jovanovich, 1992), 95. Elin McCoy, "Is School Necessary?" *Parents* 57 (April 1982): 144. Devins, "State Regulation of Home Instruction," 161. "When Parents Ask: Who Needs School?" *U.S. News and World Report* (September 22, 1980): 47. Klicka, *Home School Heroes*, 138. Sam Illis, "Schooling Kids at Home," *Time Magazine* (October 22, 1990): 84.
12. McCoy, "Is School Necessary?" 142. Farenga, *Growing without Schooling*, 204-205.
13. Moore and Moore, *Home Grown Kids*, 225. Mark and Helen Hegener, "Homeschooling Freedoms at Risk," *Home Education Magazine* 8, no. 3 (May/June 1991): 29. McCoy, "Is School Necessary?" 142.
14. Andrew Sandin, "A Comprehensive Faith," *Chalcedon Report* 363 (July 1996): 3. Somerville, "Politics of Survival." Pat Montgomery, "No Need to Compromise," *Growing without Schooling* 118 (September/October 1997): 32.
15. Tobak and Zirkel, "Home Instruction," 36-38. Richardson and Zirkel, "Home Schooling Law," 186-187. Whitehead and Crow, *Home Education: Rights and Reasons*, 287-292.

16. Richardson and Zirkel, "Home Schooling Law," 193–196. Klicka, *Homeschool Heroes*, 191.
17. John Holt, "Our Legal Situation," *Growing without Schooling* 32 (April 1, 1983): 9. James G. Cibulka, "State Regulation of Home Schooling: A Policy Analysis" in Jane Van Galen and Mary Anne Pitman, eds., *Home Schooling: Political, Historical, and Pedagogical Perspectives* (Norwood, NJ: Ablex, 1991), 104. Peterson, "Home Education v. Compulsory Attendance Laws," 285.
18. Klicka, *Home School Heroes*, 144–145. Stephen Greenburg, "The Legality of Private-School Homeschooling in California" (May 2000), <http://www.hsc.org/chaos/legal/greenberglegal.pdf> (July 26, 2007), 15–22. Some Christians also strenuously resist any effort to pass homeschooling statutes, claiming that such actions place government in charge of what God has ordained to be a family function. See, for example, Karl Reed, *The Bible, Homeschooling, and the Law* (n.p.: Christian Home Ministries, 1991). As this book goes to press the homeschooling world and national media outlets have been shocked by a recent decision by California's Second District Court of Appeals, which found that California law requires parents either to send their children to a full-time public or private school or to be taught at home by a teacher certified by the state of California. While it is doubtful that this decision will stand as is, it may result eventually in a more explicit law or clearer court interpretation of the extant law than what has to date existed in California. See Bob Egelko and Jill Tucker, "Homeschoolers' Setback Sends Shock Waves Through State," *San Francisco Chronicle*, 7 March 2008, sec. A.
19. Cochran, "Homeschooling Movement in the U.S.," 93–100.
20. Cochran, "Homeschooling Movement in the U.S.," 101–146.
21. Cochran, "Homeschooling Movement in the U.S.," 147–158.
22. Cochran, "Homeschooling Movement in the U.S.," 159–170.
23. Cochran, "Homeschooling Movement in the U.S.," 171–215. Narvis McPherson, Georgia Department of Education, telephone conversation with author (July 26, 2007).
24. Cibulka, "State Regulation of Home Schooling," 104. Bates, "Lobbying for the Lord," 11–13. Ann Lahrson-Fisher, *Homeschooling in Oregon: The 1998 Handbook* (Portland: Out of the Box Pub., 1998): 51–53. Dick and Dorothy Karman, "Timeline of Oregon Homeschool Freedoms," *OCEA Network* (July 2002), <http://www.oceanetwork.org/publications/article.cfm?id=41> (July 26, 2007).
25. "Home Schoolers Did Not Break Law, Says Maryland Judge," *Education Week* (May 9, 1984): 3. Manfred Smith, "A Lifelong Journey: Twenty Years of Homeschooling," *Maryland Home Education Association*, <http://www.mhea.com/features/journey.htm> (July 26, 2007).
26. Jim Gustafson, interview by Anna Blakeslee (October 17, 2003).
27. Susan Richman, *Western PA Homeschoolers* (Summer 1982), <http://www.pahomeschoolers.com/newsletter/issue2.pdf> (August 7, 2007).

Susan Richman, "RE: Figure?" (August 7, 2007), personal e-mail (August 8, 2007).
28. Reed, *The Bible, Homeschooling, and the Law*, 95–96.
29. Reed, *The Bible, Homeschooling, and the Law*, 97. Howard Richman, *Story of a Bill: Legalizing Homeschooling in Pennsylvania* (Kittanning, PA: Pennsylvania Homeschoolers, 1989), 13.
30. Klicka, *Home School Heroes*, 111–113. Richman, *Story of a Bill*, 33–136.
31. Reed, *The Bible, Homeschooling, and the Law*, 105. Sharon Lerew, interview by Anna Blakeslee (October 24, 2003).
32. Richman, *Story of a Bill*, 25, 28. Christine Scheller, "The Little School in the Living Room Grows Up," *Christianity Today* 46, no. 10 (September 9, 2002): 46. Kim Huber, "RE: Web Contact Form: CHAP history" (December 3, 2003), e-mail to Anna Blakeslee (August 8, 2007). Briah Doherty, "Homeschool Revolt," *Reason* 36, no. 8 (January 2005): 11.
33. Whitehead and Crow, *Home Education: Rights and Reasons*, 370. Tim Lambert, "A Home School History Lesson," *Texas Home School Coalition* (Spring 1997), <http://www.thsc.org/about_us/hs_history_lesson.asp> (August 8, 2007).
34. Perry A. Zirkel, "Home Sweet...School," *Phi Delta Kappan* 76, no. 4 (1994): 332.
35. Zirkel, "Home Sweet...School," 332. Evan Todd Yeager, "A Study of Cooperation between Home Schools and Public and Private Schools" (Ed.D. diss., Texas A&M University, 1999). Somerville, "Legal Rights for Homeschool Families," 136. Larry and Susan Kaseman, "HSLDA's 'History' Erodes the Foundations of Our Freedoms," *Home Education Magazine* (September–October 2001), <http://www.homeedmag.com/HEM/185/sotch.html> (August 8, 2007).
36. James C. Carper, "Homeschooling *Redivivus*: Accommodating the Anabaptists of American Education" in James C. Carper and Thomas C. Hunt, *The Dissenting Tradition in American Education*, 245–255. Zan Peters Tyler and James C. Carper, "From Confrontation to Accommodation: Home Schooling in South Carolina," *Peabody Journal of Education* 75, nos. 1, 2 (2000): 32–48. Zan Tyler, "My Father's Coattails," *Lifeway: Biblical Solutions for Life*, <http://www.lifeway.com/lwc/article_main_page/0%2C1703%2CA%25253D153778%252526M%25253D200871%2C00.html?> (August 8, 2007).
37. Tyler and Carper, "From Confrontation to Accommodation."
38. James Carper, "Re: South Carolina Homeschooling" (July 10, 2007), personal e-mail (August 8, 2007). "Legal Options in South Carolina," *SC Home Educators Association.org* (March 2, 2007), <http://www.schomeeducatorsassociation.org/accountability.htm> (August 8, 2007).
39. Peter Schmidt, "Measure on Home and Church Schools Fails in Iowa," *Education Week* (May 17, 1989): 14. Somerville, "Legal Rights for Homeschool Families," 137–138. Klicka, *Home School Heroes*, 73–80. Lori Higgins, "Michigan Asks Little of Teaching Parents," *Detroit Free*

Press (February 19, 2002): sec. A. Seaborne, "Does HSLDA Work with State Associations."
40. Mary McConnell, review of *The Ultimate Guide to Homeschooling*, by Debra Bell, *Journal of Law and Religion* 16, no. 2 (2001): 471. Klicka, *Home School Heroes*, 1, 216. Stevens, "Normalization of Homeschooling," 94. Bates, "Lobbying for the Lord," 3–5. Cibulka, "State Regulation of Home Schooling: A Policy Analysis," 118.
41. Klicka, *Home School Heroes*, 45. Stevens, *Kingdom of Children*, 33.
42. John Holt, *Teach Your Own: A Hopeful Path for Education* (New York: Delacorte Press, 1981), 360. Michael W. Apple, "The Cultural Politics of Home Schooling," *Peabody Journal of Education* 75, nos. 1, 2 (2000): 256. Lyman, *The Homeschooling Revolution*, 84–92. Deborah Wuehler, "Homeschooling Pioneers, Part I," *The Old Schoolhouse Magazine* (Summer 2004), <http://www.thehomeschoolmagazine.com/How_To_Homeschool/articles/411.php> (July 27, 2007).
43. Isabel Lyman, "Keeping Homeschooling Private," *The New American* (September 8, 2003): 27. Stevens, "Normalization of Homeschooling," 94.

Chapter Eight Homeschooling and the Return of Domestic Education, 1998–2008

1. Chris Strauss, "Cheyenne Kimball Sounds Off," *People* (July 24, 2006): 41. Kate Stinchfield, "Bethany Hamilton: Update," *Time* (August 7, 2006): 23. Kristen Kemp, "Girl Genius," *Girl's Life* (September 30, 1997): 16.
2. Danyel Smith, "Crazy in Love," *Essence* (February 2005): 136. "5 Reasons Why John Legend is No Ordinary Pop Star," *People* (November 6, 2006): 109–110. Margeaux Watson, "Spotlight: John Legend," *Entertainment Weekly* (November 10, 2006): 23–24.
3. "Home Works," *Wall Street Journal* (June 6, 2000): sec. A. Elizabeth Gudrais, "Homeschoolers Brush off Criticism," *Providence Journal* (September 26, 2005), <http://www.projo.com/news/content/projo_20050926_hsbee.7af3ddc.html> (August 10, 2007). "First Girl in 17 Years Wins Geography Bee Title," *ABC News.com* (May 25, 2007), <http://www.abcnews.go.com/US/story?id=3204969> (August 10, 2007).
4. Kurt J. Bauman, "Home Schooling in the United States: Trends and Characteristics," *U.S. Census Bureau* (August 2001), <http://www.census.gov/population/www/documentation/twps0053.html> (10 August 2007). Hayden, *Building Suburbia*, 4–5, 185–191, 216.
5. Daniel Princiotta and Stacey Bielick, "Homeschooling in the United States: 2003" (Washington, D.C.: National Center For Educational Statistics, 2006). Brian D. Ray, "Research Facts on Homeschooling: General Facts and Trends," *National Home Education Research*

Institute (July 10, 2006), <http://www.nheri.org/content/view/199/> (August 10, 2007). Daniel de Vise, "Schooling Has Grown Well Beyond Home: As Ranks of Parent Educators Increase, So Do Socializing, Use of Community," *The Washington Post* (March 7, 2005): sec. B. John Creason, "Home Education in Pennsylvania," *Pennsylvania Department of Education* (February 2006), <http://www.pde.state.pa.us/k12statistics/lib/k12statistics/homeEd0405withFig2.pdf> (August 10, 2007).

6. Theodore C. Wagenaar, "What Characterizes Home Schoolers? A National Study," *Education* 117, no. 3 (Spring 1997): 440–444. Daniel Princiotta, Stacey Bielick, and Chris Chapman, "1.1 Million Homeschooled Students in the United States in 2003" (Washington, D.C.: National Center For Educational Statistics, 2004). Patrick Basham, "Home Schooling: From the Extreme to the Mainstream," *Public Policy Sources* 51 (2001): 3–18. Guillermo Montes, "Do Parental Reasons to Homeschool Vary by Grade? Evidence from the National Household Education Survey," *Home School Researcher* 16, no. 4 (2006): 11–17.

7. Karen Rogers Holinga, "The Cycle of Transformation in Home School Families Over Time," (Ph.D. diss, Ohio State University, 1999). Patricia M. Lines, *Support for Home-Based Education: Pioneering Partnerships between Public Schools and Families Who Instruct their Children at Home* (Eugene, OR: ERIC Clearninghouse, 2003). Robert G. Houston, Jr. and Eugenia F. Toma, "Home Schooling: An Alternative School Choice," *Southern Economic Journal* 69, no. 4 (April 2003): 920–935. Christa L. Green and Kathleen V. Hoover-Dempsey, "Why Do Parents Homeschool? A Systematic Examination of Parental Involvement," *Education and Urban Society* 39, no. 2 (2007): 264–285. Bauman, "Home Schooling in the United States." Cathy Duffy, review of *Kingdom of Children: Culture and Controversy in the Homeschooling Movement*, by Mitchell L. Stevens, <http://www.cathyduffyreviews.com/general-book-reviews/kingdom-of-children.htm> (August 13, 2007). Ed Collom, "The Ins and Outs of Homeschooling: The Determinants of Parental Motivations and Student Achievement," *Education and Urban Society* 37, no. 3 (May 2005): 309.

8. Angie Kiesling, "The Quiet 'Boom' Market," *Publishers Weekly* (July 9, 2001): 23–25. Deborah Wuehler, "Hello and Welcome!" *The Old Schoolhouse: The Magazine for Homeschool Families*, <http://www.thehomeschoolmagazine.com/About_Us.php> (August 13, 2007).

9. Kiesling, "Quiet 'Boom' Market," 24. Angie Kiesling, "Why Johnny Learns at Home," *Publishers Weekly* (August 16, 2004): 25–26.

10. Kiesling, "Why Johnny Learns at Home," 25. Sheila Moss, "RE: B & H" (December 21, 2006), personal e-mail (August 13, 2007). "Used Curriculum Vendors," *Homeschool Resource Guide* (December 2, 2006), <http://members.cox.net/ct-homeschool/guide.htm#used> (August 15, 2007). Ann Slattery, "In a Class of Their Own," *School Library Journal* 51, no. 8 (August 2005): 44–47.

11. Kiesling, "Why Johnny Learns at home," 25–26. John Holzmann, "RE: A History of Homeschooling" (July 22, 2007), personal e-mail (August 13, 2007). Susan Richman, "RE: Figure?" (August 7, 2007), personal e-mail (August 13, 2007).
12. Lisa Guidry, "Remembering Our Roots," *Reliable Answers-News and Commentary* (May 26, 2005), <http://reliableanswers.com/hs/remembering_our_roots.asp> (August 13, 2007). McDannell, "Creating the Christian Home," 196–198. Klicka, *Home School Heroes*, 20. Christian Smith, *American Evangelicalism: Embattled and Thriving* (Chicago: University of Chicago Press, 1998). Marsden, *Fundamentalism and American Culture* (2006), 255.
13. Seelhoff, "Homeschooler's History, Part VI," 3–5. "What Are Generation Joshua and HSLDA PAC?" *Home School Court Report* 21, no. 1 (January/February 2005). Brittany Barden, "Generation Joshua—A Call to Liberty," *C.H.E.K. News* (March 2006): 1, 9, 12. Elizabeth Smith, "Homeschooling the Silent Revival," MP3 available for download at *Christian Homeschool Association of Pennsylvania* (2007), <http://www.chapboard.org/> (August 13, 2007).
14. Hannah Rosin, "God and Country: Annals of Education," *The New Yorker* (June 27, 2005): 44–49. Naomi Schaefer Riley, "The Press and Patrick Henry College," *Chronicle of Higher Education* (July 14, 2006): sec. B. "The Departed: Faculty Turnover at Patrick Henry College," *Concerned Alumni of Patrick Henry* (April 2007), <http://www.concernedalumniofpatrickhenry.com/facultydepartures.html> (August 13, 2007).
15. Rosin, "God and Country," 45–49. "God's Next Army," *Channel 4 Faith and Belief* (June 2006), <http://www.channel4.com/culture/microsites/C/can_you_believe_it/debates/godsarmy.html> (August 14, 2007).
16. Darla Isackson, "Mom Schools and Co-ops," *Meridian: The Place Where Latter-Day Saints Gather* (2004), <http://www.meridianmagazine.com/ideas/050422schools.html> (August 14, 2007). Pamela A. Vaughan, "Case Studies of Homeschool Cooperatives in Southern New Jersey" (Ph.D. diss., Widener University, 2003). Daniel Robb, "Don't Call it School," *Teacher Magazine* 18, no. 3 (November/December 2006): 24–31.
17. Laurie Goodstein, "Evangelicals Fear the Loss of their Teenagers," *New York Times* (October 6, 2006): sec. A. Margaret Talbot, "A Mighty Fortress," *New York Times Magazine* (February 27, 2000): 36. Chris Lubienski, "Whither the Common Good? A Critique of Home Schooling," *Peabody Journal of Education* 75, nos. 1 and 2 (2000): 210. "Dr. Dobson Tells Christians to Get Kids out of California Public Schools," *Homeschool World* (2002), <http://www.home-school.com/news/drdobson.html> (August 14, 2007). Associated Press, "Southern Baptist Convention Won't Support Public School Pullout Plan,"

FoxNews.com (June 14, 2006), <http://www.foxnews.com/story/0,2933,199440,00.html> (August 14, 2007).
18. Peter Beinart, "Home (School) Improvement," *Time Magazine* (October 26, 1998): 6. Mary Ann Zehr, "More Home Schoolers Taking Advanced Placement Tests," *Education Week* (26 April 2006): 12. Richman, "RE: Figure." Andrew Lawrence, "Out at Home? Barred from Sports, Some Homeschooled Kids are Fighting Back," *Sports Illustrated* (October 30, 2006): 40. Flynn McRoberts, "The Economics of Karate," *Newsweek* (November 6, 2000): 62.
19. Paul Jones and Gene Gloeckner, "Perceptions of and Attitudes Toward Homeschooled Students," *Journal of College Admission* 185 (Fall 2004): 19. Richard Joseph Bamo, "The Selection Process and Performance of Former Homeschooled Students at Pennsylvania's Four-Year Colleges and Universities," (Ed.D. diss., Lehigh University, 2003). Sean Callaway, "Unintended Admissions Consequences of Federal Aid for Homeschoolers," *Journal of College Admission* 185 (Fall 2004): 22–28.
20. Beinart, "Home (School) Improvement," 6.
21. "Accreditation and History," *Florida Virtual School* (2006), <http://www.flvs.net/general/accreditation_information.php> (August 14, 2007). National Forum on Education-Statistics, *Forum Guide to Elementary/Secondary Virtual Education* (Washington, D.C.: U.S. Dept. of Education, 2006), 1–2.
22. Luis A. Huerta, María-Fernanda González, and Chad D'Entremont, "Cyber and Home School Charter Schools: Adopting Policy to New Forms of Public Schooling," *Peabody Journal of Education* 81, no. 1 (2006): 103–139.
23. Andrew J. Rotherham, "Virtual Schools, Real Innovation," *New York Times* (April 7, 2006): sec. Op/Ed. Huerta, "Cyber and Home School Charter Schools," 124–135. Elanor Chute, "CyberSchool Begets an Education Empire," *Pittsburgh Post-Gazette* (May 30, 2006): sec. B. Joe Smydo, "Hearings to Focus on Cyber Charter Schools," *Pittsburgh Post-Gazette* (July 31, 2007): sec. B.
24. Corey Murray, "States Grapple with Virtual School Legislation," *eSchool News* (May 22, 2003), <http://www.eschoolnews.com/news/showstory.cfm?ArticleID=4419> (August 14, 2007). Scott Somerville, "2004 American Educational Research Association" (March 8, 2004), <http://www.nhen.org/forum/topic.asp?TOPIC_ID=614&whichpage=10> (August 15, 2007). Klicka, *Home School Heroes*, 279.
25. "We Stand for Homeschooling" (2003), <http://westandforhomeschooling.org/res/index.php> (August 15, 2007). Klicka, *Home School Heroes*, 275. Elizabeth Smith, "Homeschooling the Silent Revival."
26. "K Fact Sheet," *KInc.* (2007), <http://www.k12.com/press_policy/k12_fact_sheet> (August 16, 2007). Jeffry Kwitowski, "RE: K12 enrollment data" (August 22, 2007), personal e-mail (August 22, 2007). Jessica Cantelon, "Virtual Charter Schools Face Opposition from Unlikely Source," *Cybercast News Service*, CNS News.com (August 13,

2002), <http://www.cnsnews.com/ViewNation.asp?Page=%5C%5C Nation%5C%5Carchive%5C%5C200208%5C%5CNAT20020813b. html> (August 16, 2007).
27. Klicka, *Home School Heroes*, 275. John Holzmann, "RE: A History of Homeschooling" (July 24, 2007), personal e-mail (August 16, 2007). Carol Klein, *Virtual Charter Schools and Home Schooling* (Youngstown: Cambria Press, 2006), 91–95, 115–122.
28. Klein, *Virtual Charter Schools and Home Schooling*, xxiii. "Browse Groups about Home Schooling," *Yahoo!Groups*, <http://dir.groups.yahoo.com/dir/Schools—Education/Theory_and_Methods/Home_Schooling> (August 16, 2007). These figures are approximations based upon searches done on the Yahoo!Groups site. Since some group descriptions may use different terminology than what I searched for, I have likely missed many groups. Group membership, it should be noted, also ranges widely. Some groups may have only a handful of members, others several hundred.
29. Richard Rothstein, *Class and Schools: Using Social, Economic, and Educational Reform to Close the Black-White Achievement Gap* (Washington, D.C.: Economic Policy Institute, 2004). W.E.B. DuBois, *The Souls of Black Folk* (New York: Dover, 1994), 5. Grace Llewellyn, *Freedom Challenge: African American Homeschoolers* (Eugene, OR: Lowry House Publishers, 1996), 78–79, 72.
30. Llewellyn, *Freedom Challenge*, 267–288. Leslie Fulbright, "Blacks Take Education into Their Own Hands. New Ground: Once Dominated by Whites, Homeschooling Appeals to More African Americans," *San Francisco Chronicle* (September 25, 2006): sec. A. Susan A. McDowell, Annette R. Sanchez, and Susan S. Jones, "Participation and Perception: Looking at Home Schooling Through a Multicultural Lens," *Peabody Journal of Education* 75, nos. 1, 2 (2000): 124–146. Marsden, *Fundamentalism and American Culture*(2006), 327.
31. Basham, "Home Schooling: From the Extreme to the Mainstream," 7–8. Princiotta and Bielick, "Homeschooling in the United States," 5. Lyman, *The Homeschooling Revolution*, 44. Greg Beato, "The House the Burgees Built," *Reason* 36, no. 11 (April 2005): 38–39. "The New Pioneers: Black Home Schoolers," *Home School Court Report* 17, no. 4 (July/August 2001). Mary Ann Zehr, "Black Home Schoolers Share Ideas at Group's 4 Annual Symposium," *Education Week* (August 10, 2005): 20. Diamond, *Not by Politics Alone*, 241.
32. Fulbright, "Blacks Take Education Into Their Own Hands." Rebecca Catalanello, "Homeschooling: It's Not What You Think," *St. Petersburg Times* (June 26, 2005): sec. A. "Bad News for White Supremacists: Home-Schooled Blacks Do Just as Well as Homeschooled Whites on Standardized Tests," *Journal of Blacks in Higher Education* (July 31, 2000): 53–54. Jennifer James, "Why Black Children Benefit from Homeschooling," *Suite 101* (October 30, 2004), <http://www.suite101.com/article.cfm/african_american_homeschooling/111986> (August 17, 2007). Nafisa M. Rachid, "Homeschooling: It's a Growing Trend

Among Blacks," *Network Journal* 12, no. 4 (February 2005): 10. "Black Support Groups," *National African-American Homeschoolers Alliance*, <http://www.naaha.com/support.htm> (August 17, 2007). "More Black Families Home Schooling," *Jacksonville Free Press* (December 15–21, 2005): 12. "FUNgasa: Free Oneself!" *African-American Unschooling*, <http://www.afamunschool.com/fungasa.html> (August 17, 2007).

33. Gina Rozon, "Interview with Misty Dawn Thomas," *Home Education Magazine*, 17, no. 3 (May–June 2000): 20–21. Sharyn Robbins-Kennedy, "Native Americans for Home Education," <http://www.geocities.com/nuwahti/NAHE.html> (August 17, 2007). Elaine Kelsey Harvey, "High Hopes: A Qualitative Study of Family Values Guiding 'Home Schooling' Education in the Multi-Ethnic Environment of Hawaii" (Ed.D. diss., University of Hawaii, 1999). Rochelle Eisenberg, "Class Half Full," *Baltimore Jewish Times* (February 2, 2007): 36–42. Ted Siefer, "Homeschooling Growing in Popularity Among Jewish Parents, Children," *The Jewish Advocate* (August 19–15, 2005): sec. B. "25 Years of Experience in Helping Parents," *Seton Home Study School*, <http://www.setonhome.org/history.php> (August 17, 2007). Amaris Elliott-Engel, "More Muslims Teach Children in the Home-Faith Lessons Gain in Minority Group," *The Washington Times* (July 29, 2002): sec. A. Maralee Mayberry, "Teaching for the New Age: A Study of New Age Families Who Educate Their Children at Home," *Home School Researcher* 5, no. 3 (1989): 16. Kristin Madden, *Pagan Homeschooling: A Guide to Adding Spirituality to Your Child's Education* (Niceville, FL: Spilled Candy Books, 2002). "Education and Homeschooling: Preparing Ourselves and Our Children," *Stormfront White Nationalist Community*, <http://www.stormfront.org/forum/forumdisplay.php/education-and-homeschooling-28.html> (August 17, 2007). Tiny Aurora, "Elective Home Education and Special Educational Needs," *Journal of Research in Special Educational Needs* 6, no. 1 (March 2006): 55–66.

34. Susan Saulny, "The Gilded Age of Home Schooling," *New York Times* (June 5, 2006): sec. A. Michelle Conlin, "Meet My Teachers: Mom and Dad," *Business Week* (February 20, 2006): 80–81.

35. Matt Higgins, "For New-Sport Athletes, High School Finishes 2," *New York Times* (September 20, 2006): sec. Sports. Seema Mehta, "New Role for Home Study: Honing a Gift," *Los Angeles Times* (July 24, 2005): sec. B. Ruben Nepales, "Vanessa Hudgens: 'I Love Being a Filipina,'" *Asian Journal Online* (August 9, 2007), <http://www.asianjournal.com/?c=188&a=22136> (August 27, 2007).

36. Pam Sorooshian, "2004 American Educational Research Association" (February 22, 2004), <http://www.nhen.org/forum/topic.asp?TOPIC_ID=614&whichpage=2> (August 17, 2007).

37. McDannell, "Creating the Christian Home," 210–211. Judith Stacey, *Brave New Families: Stories of Domestic Upheaval in Late Twentieth Century America* (New York: Basic Books, 1990), 145. Stevens, *Kingdom of Children*, 15–16. Elizabeth S. Kaziunas, "Mother as Teacher: A Study

of Homeschooled Mothers and Dissent" (Honors thesis, Macalester College, 2003).
38. Lears, *No Place of Grace*, 57. Carlson, *The "American Way,"* 169. Allan Carlson, *Fractured Generations: Crafting a Family Policy for Twenty-First Century America* (New Brunswick, NJ: Transaction Publishers, 2005). Carper, "Homeschooling *Redivivus*," 260. Phillip Longman, *The Empty Cradle: How Falling Birthrates Threaten World Prosperity and What to Do About It* (New York: Basic Books, 2004). Joseph Pearce, *Small is Still Beautiful: Economics as if Families Mattered* (Wilmington: ISI Books, 2006). Pippa Norris and Ronald Inglehart, *Sacred and Secular: Religion and Politics Worldwide* (Cambridge: Cambridge University Press, 2004), 22–23.
39. Zygmunt Bauman, *Liquid Modernity* (Cambridge: Polity Press, 2000). Pamela Ann Moss, "Benedictines without Monasteries: Homeschoolers and the Contradictions of Community" (Ph.D. diss., Cornell University, 1995). Klicka, *Home School Heroes*, 51. Michael Apple, *Educating the "Right" Way: Markets, Standards, God, and Inequality* (New York: Routledge, 2006). Rob Reich, "The Civic Perils of Homeschooling," *Educational Leadership* 59, no. 7 (April 2002): 56–59. Lubienski, "Whither the Common Good? A Critique of Home Schooling," 207–232. Robert B. Reich, "The Secession of the Successful," *New York Times Magazine* (January 20, 1991): 16. Irving H. Buchen, *The Future of the American School System* (Lanham, MD: Scarecrow Education, 2004). Robert Sanborn, Adolfo Santos, Alexandra L. Montgomery, and James B. Caruthers, "Four Scenarios for the Future of Education" *Futurist* 38, no. 7 (January/February 2005): 26–30. Chris Anderson, *The Long Tail: Why the Future of Business Is Selling Less of More* (New York: Hyperion, 2006). Daniel Pink, *Free Agent Nation: How America's New Independent Workers are Transforming the Way We Live* (New York: Warner Books, 2001).
40. Carper, "Homeschooling *Redivivus*," 260.

Index

100 Top Picks for Homeschool Curriculum (Duffy), 155
4-H clubs, 63

A Beka Book, 153–54, 216
abortion, 106, 253
Accelerated Christian Education (ACE), 152–53, 185, 187
Adams, Abigail, 27
Adams, Henry, 27–8
Adams, John Quincy, 27–8
Addams, Jane, 79
Adventism, Seventh Day, 77, 127–33, 144, 149, 166
advice literature, 36–7, 71–2
African-Americans, 23–5, 34, 48, 53–5, 188, 219–22. See also homeschooling—minorities and
Ahmanson, Howard Jr., 136–37, 158
American Association of Christian Schools (AACS), 109, 112
American Enterprise Institute, 104
American Eugenics Society, 63
American Homeschool Association (AHA), 169
Appalachian Bible Company, 206
Apple, Michael, 198–99, 225
apprenticeship, 20–2, 79–80
Armey, Dick, 168
Arminianism, 8, 30
Arons, Stephen, 120
The Art of Education (Dobson), 126
Asbury, Francis, 31
Association of Christian Schools International (ASCI), 108

Awakening, Great, *see* Great Awakening

Baby and Child Care (Spock), 90
Bailey, Liberty Hyde, 63
Baker, David, 67–8
Ball, William, 146
Baruch, Dorothy, 65–6
Bates, Stephen, 198
Bauman, Zygmunt, 225
Beatty, Barbara, 36, 65
Beecher, Catharine, 34, 37
Beinart, Peter, 212
Berle, A. A., 77
Better Late than Early: A New Approach to Your Child's Education (Moore and Moore), 131
Bible
    early American literacy and, 17–18, 21
    family and, 12–14, 35, 59, 75
    in public schools, 42, 69, 107
    Reconstructionism and, 135–39
    slavery and, 48–9, 53–4
Big Book of Home Learning (Pride), 155
Bird, Wendell, 159
birth rate, 10–11, 33, 55–6, 62–3, 225–26
Blankenship v. Georgia (1979), 185–86
Blor, Ella Reeve, 73
Blount v. Department of Educational and Cultural Services (1988), 177
Blumenfeld, Samuel, 194, 253

266   Index

Bob Jones University v. United States (1983), 111–12
Bob Jones University Press, 153–54, 216
Bond, Julian, 188
Bonner, John, 41
Brace, Charles Loring, 79–80
Bradford, William, 8
Braude, Anne, 31
Broadman and Holman, 206
Bronfenbrenner, Urie, 93
Brooks, David, 156
Brown v. Board of Education (1954), 87
Browne, Robert, 7
Buckley, William F. Jr., 74
Burges, Joyce and Eric, 220–21
Bushnell, Horace, 30

Calvary Chapel Movement, 147, 170
Calvert School, 77, 189–90
Calvinism, 16, 28–32, 109, 135
Cameron, Ann, 191
Cardinal Mindszenty Foundation (CMF), 105
Carlson, Allan, 224
Carper, Jim, 196–97
Carter, Jimmy, 92
Cassidy, M. A., 70
catechism, 13, 17–18
Catholicism, 17, 42–3, 60, 67–9, 80, 106, 167
Centro Intercultural de Documentación (CIDOC), 124
Chalcedon Foundation, 136, 139
children's literature, 36–7, 54
A Choice Not an Echo (Schlafly), 105
Christian Anti-Communism Crusade, 103, 105
Christian Home Educator's Curriculum Manual (Duffy), 155
Christian Home Educators of Ohio (CHEO), 171–72

Christian Homeschool Association of Pennsylvania (CHAP), 192–93
Christian Liberty Academy Satellite Schools (CLASS), 145–47, 153, 161
Christian Nurture (Bushnell), 30
Christian Schools International (CSI), 109
Christianity Today, 106
Cibulka, James, 198
Civil War, 49, 51–5
closed communion homeschooling, 143–44, 157, 169–73, 175, 192–93, 196–97, 205, 208–10, 214, 217, 219
Cob, Lyman, 37
Cohen, Rose, 72–3
cold war, 87
Colfax, David and Micki, 166
Collom, Ed, 205
Commonwealth v. Roberts (1893), 70
communes, 96–100
compulsory attendance laws, *see* public schools—compulsory laws
Comstock Act, 56, 58
conferences, *see* conventions
Connecticut Homeschool Network, 199
Connors, Diane, 199
Constitutionality of homeschooling, *see* homeschooling—legal aspects
conventions, 148–52, 154, 157, 160, 163–65, 167, 169–73, 193, 205–7, 218, 220
Cook, Harriet Campbell, 74
cooperatives, 19, 211–14
correspondence programs, 77–9
courtship, *see* natalism
Cowley, Malcom, 59–60
Cremin, Lawrence, 2
Critchlow, Donald, 106
curriculum, 44
    controversies over, 107, 118–19, 153, 166–67, 216, 218, 220

used by homeschoolers, 83–4, 133, 138–39, 145–46, 152–56, 205–7, 211–13
cybercharter schools, 204, 214–18

Dalton, Lucinda, 75
dame school, 19
Darrow, Clarence, 62
Darwinism, 54, 60, 62, 73
Day of Doom (Wigglesworth), 18
de Tocqueville, Alexis, 34, 54
deindustrialism, 224–25
DeKoven, Marianne, 96
Delconte v. State (1988), 180
Demos, John, 4
Deschooling Society (Illich), 124
Dewey, John, 72, 138
Diamond, Sara, 221
Dobson, James, 92, 132–34, 158, 160, 163, 171, 190, 212
Dobson, Linda, 111–12, 126–27, 166, 172, 244
Donahue, Phil, 126–27, 133–34, 166
Douthat, Ross, 137
Downing, Lucia, 39–40
The Drinking Gourd, 220
DuBois, W. E. B., 219
Duffy, Cathy, 154–55, 167, 205
Duro v. District Attourney (1983), 178

Edison, Nancy, 44–5
Edison, Thomas, 44–5, 59
Education Entrance Exam (EEE), 196
Edwards, Jonathan, 28
Eldredge, Tom, 191
Elson, Ruth Miller, 41
Emerson, Ralph Waldo, 35
Equal Rights Amendment (ERA), 105–6, 159
Escape from Childhood (Holt), 124

Fair Housing Act (1968), 88
Falwell, Jerry, 102, 159

Family Matters: Why Homeschooling Makes Sense (Guterson), 166
Family School (Peabody), 66
Farber, David, 94
Farenga, Patrick, 165
Farris, Michael, 2, 137, 146, 149, 159–60, 162, 168–69, 178–79, 192, 209–10
fathers, 14–16, 35, 39
Federal Housing Administration (FHA), 86–7
Feminine Mystique (Friedan), 88
feminism, 59, 63, 224–25
Fidler, Ray, 189
Filmer, Sir Robert (Filmerian outlook), 13–14
Fireside Correspondence School, 77
Fisk, Elizabeth, 76
Fleigelman, Jay, 25
Florida Virtual School (FVS), 214
Ford, Henry, 85
Forstmann, Theodore, 1
Foster, Charles, 39
Frank, Thomas, 91
Franklin, Benjamin, 11
Freedom and Beyond (Holt), 124
Friedan, Betty, 88, 104
Fugate, Richard, 167
FUNgasa, 222

Gaither, Bill and Gloria, 114
Gangel, Kenneth, 110
Generation Joshua, 209
Gentle Spirit, 156, 171–73
Gibbs, David, 146
Gibson Patricia, 186
Goldin, Claudia, 68–9
Goldwater, Barry, 105
Goodrich, Samuel, 41
Gothard, Bill, 151–52, 169, 171
Graham, Billy, 106
Great Awakening, 31–2
Green v. Connally (1970), 111
Griggs, Frederick, 77

Griggs University and International Academy, 77
Grove City v. Bell (1984), 161
Growing Without Schooling (GWS), 125–26, 156, 165–66, 186, 211
Guidry, Lisa, 208
Gustafson, Jim and Gloria, 190
Guterson, David, 166

H.R. 6, 168
Hamilton, Bethany, 201
Hanson v. Cushman (1980), 177
Harris, Gregg, 2, 147–50, 154, 157, 164, 167, 169, 171–73, 189, 193
Harris, Joe Frank, 188
Harris, Joshua, 147, 156
Hatch, Nathan, 31
Hawthorne, Nathaniel, 36
Head Start, 130
Hegener, Mark and Helen, 166, 169, 172, 244
HELP for Growing Families, 156, 172
Hemmenway, Moses, 30
Henry, Carl F. H., 106
Hewitt Foundation, 130–32
Hillyer, Virgil, 77
Hiner, N. Ray, 15
Hodge, Charles, 30
Holt, John, 2, 120, 122–28, 131, 133–34, 144–45, 156, 165–66, 176, 182, 184, 190, 198
Holzmann, John, 207
home birth, 4, 97, 100, 151–52, 171. *See also* natalism
home economics, 64–66
Home Education Magazine, 126, 166, 169, 172, 205
Home Education: Rights and Reasons (Whitehead and Bird), 159
Home Grown Kids (Moore and Moore), 132–34, 155, 190
Home Oriented Unique Schooling Experience (HOUSE), 125–26
Home School Burnout (Moore), 167
Home School Court Report, 163, 196, 221
Home School Legal Defense Association (HSLDA), 157, 159, 162–65, 167–70, 175–76, 178–79, 185, 192–93, 195–97, 208–10, 216–17, 221–22
Home Study Institute (HSI), 77, 127
homeschooling
 attitudes of school personnel toward, 181–84, 191, 194–95, 213
 cooperatives, *see* cooperatives
 curriculum, *see* curriculum
 higher education and, 213
 legal aspects of, 70–1, 119–20, 145–46, 157–64, 166, 172–73, 175–87, 189, 192, 194–97, 253
 legislative issues concerning, 168, 175, 179–80, 184–93, 195–97, 199–200, 254
 media coverage of, 126, 143, 146, 153, 163, 198–99, 246
 minorities and, 210–2, 204, 219–24
 opposition to, 181–82, 194–95, 225
Homeschooling for Excellence (Colfax), 166
Homeschooling Today, 205
house churches, *see* natalism
How Children Fail (Holt), 123
How Children Learn (Holt), 123
Howell v. State (1986), 178
Hudgens, Vanessa, 223
Huggins, Eliza and Mary, 47
Hull House, 79
Hungate, William, 93
Hutchinson, Anne, 15

Index

Illich, Ivan, 124
In Re Davis (1974), 180
In Re Franz (1977), 177
In Re Sawyer (1983), 180
industrial revolution, 56–65
Information Network for Christian Homes (INCH), 197
Ingersoll, Robert, 73
Ingram, Carol, 169
Inherit the Wind, 62
Instead of Education: Ways to Help People Do Things Better (Holt), 125
Institutes of Biblical Law (Rushdoony), 136, 138
Intellectual Schizophrenia (Rushdoony), 137
International Correspondence Schools (ICS), 78
Internet, 4–5, 169–70, 172–73, 207–8, 225
Interstate Highway Act (1956)

James, Jennifer, 221
Jeffrey v. O'Donnell (1988), 183, 192
Jernigan v. State (1981), 178, 184
John Birch Society, 103–4
Jorgensen, Theodore, 75–6

K^{12} (curriculum), 217
Kaestle, Carl, 38
Kaseman, Larry and Susan, 166, 195
Katz, Lawrence, 68–9
Kearney, Maeghan and Michael, 201
Keller, Morton, 64
Kelly, Florence, 63–4
Kimball, Cheyenne, 201
The Kindergarten Messenger (Peabody), 66
King, Martin Luther Jr., 117
Kinmont, Joyce, 127, 211

Klicka, Chris, 146, 160–63, 173, 178, 184, 197, 208, 213, 217, 225
Klingberg, Frank, 55
Knox v. O'Brien (1950), 183
Kozol, Jonathan, 121
Kruschev, Nikita, 87

Ladies' Home Journal, 58
LaHaye, Beverley, 159–60
LaHaye, Timothy, 121, 136, 160, 209
Larcom, Lucy, 40
Lassonde, Steven, 67
Late Great Planet Earth (Lindsey), 93, 158
Lears, Jackson, 62
Leeper v. Arlington Independent School District (1987), 180, 194–95
Legend, John, 201–2
Levin, Charles, 178
Levisen, Marjorie, 128–29
Levitt, Abraham, 87
Lindsey, Cheryl, 156, 170–73
Lindstrom, Paul, 145–46, 161
Lines, Patricia, 142
Locke, John, 14, 19, 24–5, 35, 44
Long, Huey, 19
Lubienski, Chris, 225
Lucas, Eliza, 24–5
Lyman, Isabel, 199

Mann, Horace, 1, 38, 44, 138
Marin, Peter, 121
marriage, 10–12, 89, 96–7
Marsden, George, 136, 220
Maryland Home Education Association (MHEA), 189
maternalism, 63, 66
Mather, Cotton, 18, 36
Mather, Increase, 15, 18–19
McDannell, Colleen, 104, 224
McDonald, Cornelia Peake, 51–3
McGuffey's Readers, 54
Mead, Margaret, 93

Menken, H. L., 62
Messerli, Jonathan, 44
The Messianic Character of American Education (Rushdoony), 137–38
Meyer v. Nebraska (1923), 177
Meyerowitz, Joanne, 89
Michigan v. Bennett (1993), 178
Miller, George, 168
Milliken v. Bradley (1974), 88
Minutemen, 103
Moen, Matthew, 163
Mom School, 211
Montes, Guillermo, 204
Montgomery, Pat, 166, 182, 197
Moore, Dorothy, 2, 128–34, 160, 163, 167, 182, 190
Moore, E. Ray, 212
Moore, Raymond, 2, 128–34, 144–45, 147, 148–49, 160, 163, 166–67, 169, 182, 186–87, 190, 193–94
Moore Academy, 133
Moral Majority, 102, 159
Moran, Gerald, 16
Morgan, Edmund, 20
Mormonism, 74, 117–18, 127, 133, 167, 186
Moss, Pamela Ann, 225
Murphy v. State (1988), 177
Murray, Charles J., 194

Nagel, Ed, 183
Nash, Margaret, 43
Natalism, 4, 97, 100, 151–52, 155–56, 171, 210
National African-American Homeschoolers Alliance, 221
National Black Home Educators Resource Association, 221
National Center for Home Education (NCHE), 169, 208
National Congress of Mothers (National Congress of Parents and Teachers), 69
National Council of Parent-Educators, 187

National Education Association (NEA), 130, 181
National Home Education Legal Defense (NHELD), 168
National Home Education Network (NHEN), 169
National Home Education Research Institute (NHERI), 221
National Home Study Council (NHSC), 78
National Homeschoolers Association (NHA), 169
Native Americans, 9–10, 47, 222
Network of Black Homeschoolers, 220
New Deal, 63–4
New England Primer, 30
New York Children's Aid Society (NYCAS), 79–80
Nichols, Steve and Ann, 187
Nichols-White, Donna, 219–20
Nixon, Richard, 87
No Greater Joy Ministries, 199
Nogaki, Katy, 223
Noll, Mark, 54
Norman, Larry, 101
North, Gary, 145, 161
North Star (homeschool cooperative), 211
Norton, Mary Beth, 13–14

O. W. Coburn School of Law, 162
Old Schoolhouse Magazine, 205–6
open communion homeschooling, 143–44, 157, 165–70, 175, 192–93, 196–97, 205, 217, 219
Oral Roberts University, 162
Oregon Christian Home Education Association Network (OCEAN), 189
original sin, 22–3, 28–32, 113, 138
orphans, 22, 46–7, 52, 79–80
Ownby, Ted, 35

PA Homeschoolers, 190, 207

Packard, Ron, 217
Padgett, Wimbric and Marion, 187
Palmetto Muslim Homeschool Resource Network, 222
Parent Teacher Organizations (PTOs), 69–70
parenting manuals, *see* advice literature
Parr v. State (1927), 70
Patrick Henry College (PHC), 209–10
Peabody, Elizabeth, 66
Peale, Norman Vincent, 91
Pearl, Michael and Debi, 199
Pedrick, Mary, 61
Penn, William, 19
Pennsylvania Cyber Charter School, 215
Penny, Amy, 53
People in Interest of D. B. (1988), 180
People v. Darrah and Black (1986), 185
People v. DeJonge (1993), 178
People v. Levisen (1950), 180
People v. Turner (1953), 180
Perchemlides, Peter and Susan, 118–19
Perry, Arthur, 70
Pestalozzi, Johan, 37
petty school, *see* dame school
Phelps, William, 60
The Philosophy of the Christian Curriculum (Rushdoony), 138–39, 145
Pierce v. Society of Sisters (1925), 70–1, 177
Practical Homeschooling, 155, 157, 205
Pratt, Caroline, 66
Pride, Mary, 2, 154–56, 167, 172–73
private schools, 43, 108–12, 138–39, 181, 187
Professional Tutors of America, 222
Protestantism, 7–8, 17, 28–32, 36, 53–4, 101–2, 106, 113–14, 136, 208, 224
    and homeschooling, 142–45
    and public education, 38–9
public schools, 38–43, 66–70, 76, 85
    criticism of, 121–22, 127, 137–38, 159
    compulsory laws, 63, 67, 70–1, 89, 179–80
putting out, *see* apprenticeship

Ray, Brian, 204, 220–21
reading, 18, 48, 57–9, 72–3
Reagan, Ronald, 93
Reconstructionism, 135–39, 145–46, 152–55, 157–62, 188, 194
Reich, Rob, 225
Reisman, David, 72, 91
Regent University, 162
Richardson, Samuel, 36
Richman, Howard and Susan, 190–92, 207, 213
Robinson, John, 8, 22
Roe v. Wade (1973), 94, 106, 253
Roemhild, Terry and Vicki, 186–87
Roemhild v. State (1983), 180, 183, 186–87
Roman Catholicism, *see* Catholicism
Roosevelt, Theodore, 62–3
Root, George Frederick, 53
Rose, Patrick, 216
Rousseau, Jean-Jacques, 19, 35
Rushdoony, Rousas J., 2, 134–40, 145–46, 157–58, 161, 187, 194
Rutherford Institute, 158–59, 162, 250

Saleem, Fatima, 222
Salmon, Marylynn, 15
Schaeffer, Francis, 136, 158
Schickel, Norbert and Marion, 83–5
Schlafly, Phyllis, 92, 104–7, 133, 187
School Can Wait (Moore and Moore), 132

Schwarz, Fred, 103, 105
Scoma v Chicago Board of Education (1974), 177–78, 180
Scopes, John, 62
The Second American Revolution (Whitehead), 158
Second Great Awakening, see Great Awakening
secular humanism, 108, 121, 136, 160
Seelhoff, Cheryl Lindsey, see Lindsey, Cheryl
Seelhoff v. Welch (1998), 172–73
selectmen, 12–13
Self Culture (Berle), 77
Sellers, Charles, 32
The Separation Illusion: A Lawyer Examines the First Amendment (Whirehead), 158
Sessions, Bob and Linda, 126
Seton Home Study School, 222
settlement houses, 79
Seventh Day Adventistism, see Adventism, Seventh Day
Sewall, Samuel, 18, 23
Seymour, Bennett, 75
Sharlet, Jeff, 135
Sharpe, Shelby, 194–95
Shaw, Connie, 186
Sheffer, Susannah, 165
Sigourney, Lydia, 43–4
Simpson, Carol, 216
Singer, John, 117–18, 121
Smith, Christian, 208
Smith, Daniel, 35
Smith, Elizabeth, 217
Smith, J. Michael, 160, 222
Smith, Manfred, 189–90
Smith, Timothy, 39
Smith, Will and Jada Pinkett, 201
Smyth, John, 7
Snaring, Caitlin, 202
Society to Encourage Studies at Home, 77–8
Somerville, Scott, 142, 183, 216, 253

Sorooshian, Pam, 223–24
South Carolina Association of Independent Home Schools (SCAIHS), 196–97
Spock, Dr. Benjamin, 90–2
Stacy, Robert, 210
Staples, Cameron, 199
State v. Bailey (1901), 71
State v. Counort, (1912), 70
State v. Edgington (1983), 177
State v. Hoyt (1929), 70, 180
State v. Lowry (1963), 180
State v. Massa (1967), 180, 183
State v. Melin (1988), 178
State v. Moorhead (1981), 184
State v. Newstrom (1985), 183
State v. Nobel (1980), 178
State v Patzer (1986), 178
State v. Peterman (1904), 70
State v. Popanz (1983), 183, 187
State v. Riddle (1981), 178, 184
State v. Schmidt (1987), 178
State v. Whisner (1976), 181
Stephens v. Bongart (1937), 183
Stevens, Mitchell, 114, 143, 165, 198
Stevenson, Deborah, 168
Stoddard, Samuel, 36
Stowe, Harriet Beecher, 61–2
Suarez, Paul and Gena, 205–6
suburbs, 57, 62, 85–9, 103–4, 110, 113, 203
Sunday school, 42
support groups, 141–45, 150, 157, 186–87, 204, 218–19, 222–23, 261
Swan v. Charlotte-Mecklenburg (1971), 88

Taylor, Susie King, 48
Teach Your Own (Holt), 134
The Teaching Home: A Christian Magazine for Home Educators, 157, 164, 172–73, 189, 208
Texas Education Agency v. Leeper (1994), see *Leeper v. Arlington Independent School District*

Thampy, George, 202
Thomas, M. Carey, 73
Thurmond, Strom, 195
Ticknor, Anna Eliot, 77
Tillman, Ben, 53
tithingmen, 13
Toffler, Alvin, 121
Treatise on Domestic Economy (Beecher), 37
Troen, Selwyn, 44
Trollope, Frances, 31–2
tutoring, 19–20, 24, 43–4, 46, 70, 74–5, 193, 222–23
Tuuri, Dennis, 188
Tyler, Zan, 195–97, 206

United States v. Orito (1973), 177
unschooling, 125–26, 166, 182–83, 185, 206, 222

Van Galen, Jane, 143
Van Til, Cornelius, 135
Veteran's Administration, 86
Victorianism, 56–63, 73, 79, 104
Vinovskis, Maris, 19
virtual charter schools, *see* cybercharters

Wallace, Circe, 223
Walton, S. Courtney, 222
Watts, Isaac, 30–1
The Way Home: Beyond Feminism, Back to Reality (Pride), 155

Webster, Daniel, 34
Weeks, Mary Harman, 69
Weems, "Parson" Mason Locke, 35
Welch, Robert Jr., 103–4
Welch, Sue, 157, 164, 171–73, 193
Weller, Charles, 45
Wells, Robert, 16
West, Elliott, 45, 75
Weyrich, Paul, 103, 111, 212
White, Ellen G., 128–29
Whitehead, John, 146, 158–59, 162, 250
Wiebe, Robert, 55
Wigglesworth, Michael, 18
Wilkerson, Gilbert and Gloria, 220
Wilkinson, Bruce, 148
Will Early Education Ruin Your Child? (Fugate), 167
Willard, Frances, 45–6, 59
Williams, William, 20
Winslow, Flora Davis, 41
Wisconsin v. Yoder (1972), 177–78
Women's Christian Temperance Union (WCTU), 59
Woodhouse, Susan, 48
Wright v. State (1922), 70
Wuthnow, Robert, 102

Young, Brigham, 74

Zirkel, Perry, 179

CPSIA information can be obtained at www.ICGtesting.com
Printed in the USA
LVOW12s0948021213

363508LV00002B/190/P

9 780230 606005